Power From The Wind

The 1250-kilowatt test unit at Grandpa's Knob.

Power From The Wind

PALMER COSSLETT PUTNAM

 VAN NOSTRAND REINHOLD COMPANY

New York   Cincinnati   Toronto   London   Melbourne

Van Nostrand Reinhold Company Regional Offices:
New York   Cincinnati   Chicago   Millbrae   Dallas

Van Nostrand Reinhold Company International Offices:
London   Toronto   Melbourne

Library of Congress Catalog Card Number: 74-4449
ISBN: 0-442-26650-2

Manufactured in the United States of America

Published by Van Nostrand Reinhold Company
450 West 33rd Street, New York, N.Y. 10001

Published simultaneously in Canada by Van Nostrand Reinhold Ltd.

15 14 13 12 11 10 9 8 7 6 5 4 3 2 1

Library of Congress Cataloging in Publication Data

Putnam, Palmer Cosslett, 1900-
   Power from the wind.

   1. Air-turbines. 2. Wind power. 3. Electric
power production. I. Title.
TK1541.P8 1974   621.4'5   74-4449
ISBN 0-442-26650-2

DEDICATED TO

M. T. P.

FOREWORD

In 1939 the directors of the S. Morgan Smith Company, manufacturers of hydraulic turbines, decided to explore the possibilities of large-scale wind-turbines as an additional source of power, and as a means of diversifying their product. To harness the power in the wind on a large scale required a knowledge of the habit of wind, about which science had little to say to us. To enter the field would require basic research.

T. S. Knight, Vice-President of the General Electric Company, had introduced Palmer Cosslett Putnam to us in 1939. Putnam had reviewed alternative designs of large windmills and proposed one of his own. We also reviewed the problem, and concluded that we liked Putnam's design. Encouraged by Walter Wyman, who secured the collaboration of the Central Vermont Public Service Corporation, we decided to engineer one full-scale unit of the Smith-Putnam Wind-Turbine, to determine if it was technically sound; and to undertake the necessary basic research into the habit of wind.

Under Putnam's leadership we organized a team of eminent men in the various fields of science and engineering, directed by Professor John B. Wilbur of the Massachusetts Institute of Technology, who served as Chief Engineer of the Project.

In six years of design and testing of the 175-foot, 1250-kilowatt experimental unit on Grandpa's Knob near Rutland, Vermont, in winds up to 115 miles per hour, we have satisfied ourselves that Putnam's ideas are practical, and that regulation is sufficiently smooth. We think we could now design, with confidence, 2000-kilowatt wind-turbines incorporating important improvements leading to smoother operation, simpler maintenance, and lower cost.

As soon as we learned from the test results what a production unit would look like, we began to recompute the over-all economics.

In 1939, based on 1937 prices, we had estimated that on a 4000-foot ridge in northern New England, a battery of ten 1500-kilowatt wind-turbines, 200 feet in diameter, would generate energy at about $0.0025 per kilowatt-hour.

Our 1945 estimates, based on a preproduction model similar to the test unit, showed that at the best site in Vermont the cost would be about $0.006 per kilowatt-hour. Jackson and Moreland estimated that the Central Vermont Public Service Corporation could afford to pay no more than about $125 per kilowatt for a block of wind-turbines generating energy at this cost. At 1945 prices, it would have cost us about $190 per kilowatt to install a block of such capacity. It is true that the cost savings proposed in Chapter XII would, if realized, more than bridge this gap. But we had already spent a million and a quarter dollars, and it would cost several hundred thousand dollars more to find out whether we could actually sell in this market at a profit. Our stockholders were unwilling to make the additional investment and, accordingly, we have reluctantly abandoned the project, placing the patents in the public domain.

Although we did not publicize it, the experiment on Grandpa's Knob commanded world-wide attention, and even during the war we continued to receive inquiries from many countries. Interest in wind-power is widespread. Accordingly, we have asked Putnam to summarize our explorations in this field and to evaluate the future of large-scale wind-power.

I agree with Putnam's conclusion that large wind-turbines will be limited to applications in special regions, which, in the aggregate, however, may amount to a considerable market—a market which I hope to live to see developed.

BEAUCHAMP E. SMITH

York, Pennsylvania
November 1946

PREFACE

For six years a group of eminent scientists and engineers, under the auspices of the S. Morgan Smith Company, has worked on the technical problems of the 1250-kilowatt windmill. Leading specialists in many fields have served as consultants. The ultimate economics of wind-power have been explored in a preliminary way.

This book is a summary of the problems which faced us in 1939, our attempts at solving them, our findings and conclusions.

It is directed to anyone interested in man's instinctive urge to subdue and harness his environment, and particularly to those in Government or Industry who are interested in eking out dwindling supplies of low-cost fuels with other sources of power.

Carl J. Wilcox, who wrote many of the original engineering reports, has prepared the drawings and organized the material for the text, which has been reviewed by all of our associates. To him warm thanks are due for help and criticism.

PALMER COSSLETT PUTNAM

Boston, June, 1947

vii

CONTENTS

INTRODUCTION

The great wind-turbine on a Vermont mountain proved that men could build a practical machine which would synchronously generate electricity in large quantities by means of wind-power. It proved also that the cost of electricity so produced is close to that of the more economical conventional methods. And hence it proved that at some future time homes may be illuminated and factories may be powered by this new means. Having provided these proofs, the installation has since been relinquished by the group that created it. It had served its purpose; economic factors worked against it; other responsibilities were to be met; other work was to be done.

The project that this volume discusses must not be written off with such dispatch as the preceding remarks suggest. However important were the three immediate proofs that it gave—and I regard them as highly important—this project, when viewed against a larger scene, grows still greater in significance.

The decision to undertake the design, construction, and operation of a wind-turbine to generate electricity led to a pioneering effort which aptly and fully illustrates how man utilizes intelligence in mastering his environment for the purpose of advancing his general welfare, in proceeding from ignorance to theoretical knowledge, and then reducing theoretical knowledge to practical application. Moreover, earlier harnessing of the winds to drive ships, to turn millstones, or to pump water, had resulted primarily in individual, independent applications; the Vermont project called for the correlation and integration of the installation with other installations using other types of power, and the tying of the result into a wide-flung and complex network unobtrusively serving thousands of people who knew little if anything of its existence.

Complex as are the electrical systems of which it was a part and the economic system out of which it grew, the wind-turbine is notable as the physical result of a project conceived and carried through by free enterprisers who were willing to accept the risks involved in exploring the frontiers of knowledge, in the hope of ultimate financial gain. Note that such financial gain would not have been at the cost of some other part of the economy. Large-scale wind-power will not create unemployment, for, in general, it displaces nothing, but rather draws on a new source to supplement steam and water, and can enlarge the use of the product it creates. It could, fully developed, bring light, heat, and power to regions that otherwise could not afford such services, or, in fact, because of physical difficulties, could not have them at all.

xi

INTRODUCTION

Free enterprise thus demonstrated in this project that it functions as effectively in the involved social structure of the present as it did in simpler societies in the past. So too in this project was demonstrated the ability of complex science and technology to focus a score of specialized skills on the various aspects of a problem, coordinating and collaborating in effective teamwork. That this scientific co-operation in a wide number of fields was carried out in this instance by a tem-porary organization mustered simply for the single undertaking, using people scattered from coast to coast, and nearly all on a part-time basis, is remarkable.

The project, in addition, illustrates what may be secured from a pioneering venture into new fields of knowledge. Scientific contributions of distinct im-portance resulted from it, particularly to knowledge of how winds behave in mountainous country, to knowledge of icing conditions, to knowledge of the be-havior of metals under new and exacting conditions. Wind-energy surveys in a half-dozen portions of the globe may well be a consequence of the undertaking.

The three immediate proofs which the project gave were highly important. The four long-range demonstrations which resulted from it are a further and lasting cause for its being considered of prime significance. The story is ably re-corded in this volume.

V. BUSH

December, 1946

Chapter I

PERSONALITIES AND HISTORY OF THE SMITH–PUTNAM WIND–TURBINE PROJECT, 1934–1945

In the fall of 1941 something new had been added to the generating system of the Central Vermont Public Service Corporation. Motorists in central Vermont saw, from 25 miles away, a giant windmill (frontispiece), its polished sunlit blades flashing on top of the 2000-foot Grandpa's Knob,* 12 miles west of Rutland and overlooking the Champlain Valley. This was the experimental Smith-Putnam Wind-Turbine, undergoing its first tests. The unit was rated at 1250 kilowatts, enough to light a town, and was feeding power into the utility company's system, permitting water to be stored behind the dams when the wind blew more than 17 miles an hour.

The synchronous generator, operating in phase with the other generators on the system, was driven, through gears and a hydraulic coupling, by a two-bladed stainless-steel turbine 175 feet in diameter, whose hub was at a height of 120 feet above the summit of the smoothly glaciated Knob. The turbine responded to changes in wind direction by rotation in yaw. Power output was regulated hydraulically by controlling the pitch of the blades with a speed governor which responded to changes in rotational speed of the turbine shaft. The blades were free to cone down-wind, moving like the ribs of an umbrella. For the first time, wind had been harnessed to drive a synchronous generator feeding directly into the high-line of a utility network.

The experiment is another proof that the spirit of exploration and adventure had not yet died out in those ancient citadels of capitalism, New England and Pennsylvania. This chapter briefly describes the development of the project, backed by a group of Down-east Yankees, and free enterprisers from York, Pennsylvania.

In 1934 I had built a house on Cape Cod and had found both the winds and the electric rates surprisingly high. It occurred to me that a windmill to generate alternating current might reduce the power bill, provided the power company would maintain stand-by service when the wind failed, and would also permit

---

* This peak had not been distinguished on maps by a separate name. It was bought from a Vermont farmer whose family always referred to it as "Grandpa's." Because of this, and its shape, we christened it Grandpa's Knob.

me to feed back into its system as dump power the excess energy generated by the windmill.

But when I came to compute the size of windmill needed to carry the peak load of the all-electrical house in the prevailing wind, it was clear that none of the small units, widely used for farm lighting, would be large enough. A much larger unit would be necessary.

Aaron Davis, a friend and neighbor, called attention to the design of a Finn, Savonius, who had just been commissioned by Colonel Henry Huddleston Rogers to put up three Savonius-type rotors on Colonel Rogers' estate at

Southampton, Long Island. A review of this design made it clear that it was inherently inefficient per unit of weight, since all of the area swept was occupied by metal. Two narrow rapidly moving blades extract more energy from the wind, per unit of area swept.

This study brought me into touch with Elisha Fales, who had been among the first to apply the aerodynamic lessons of the First World War to the problem of the windmill. In order to obtain a preliminary indication of the

PALMER COSSLETT PUTNAM

Manager of the Smith-Putnam Wind-Turbine Project.

strength of the wind at an average site on Cape Cod, Fales lent me one of his small two-bladed direct current test units, which, mounted on a 60-foot pipe, served briefly and intermittently as a sort of anemometer.

The results of these measurements were inconclusive, but stimulated me to survey the previous work as reported in the literature and to study the designs for large, wind-driven, direct-current and induction generators that had been proposed by Flettner, Mádaras, Kumme, Darrieus, the Russians, and Honnef (Chapter VI).

My conclusion was that, if an economically attractive solution to the problem existed, it lay in the direct generation of alternating current by a very large, two-bladed, high-speed windmill, feeding into the lines of an existing hydro, or steam and hydro, system. Thus, existing hydro-storage would provide the capacity to tide over periods of no wind. When the wind blew, the dispatcher could shut down the hydro units and accumulate additional stored water in the reservoirs. From the point of view of a hydro system the energy in the wind then became merely increased stream flow. To assure utmost wind velocity, the site should have airfoil characteristics and should be capable of speeding up the wind over its summit.

Another friend and neighbor, Harold Sawyer, made a useful suggestion, which, however, did not become technically practical for another ten years. Sawyer pro-

2

posed that there be incorporated between the wind-turbine and the generator a torque-limiting device that would be incapable of passing along to the generator a torque in excess of a predetermined value, regardless of those gust-generated power surges from the turbine, too sudden to be prevented by the speed governor. The mechanical solution proposed by Sawyer did not prove feasible, but the electric coupling now available seems most promising. After a few months Sawyer left to join the staff of the Mc-Math-Hurlburt Observatory in Michigan, assigning his interest in the project to me.

I continued to develop the design features which later became incorporated in the test unit, and made some rough cost estimates. These looked promising to Dr. Vannevar Bush, then Dean of Engineering at the Massachusetts Institute of Technology, and in 1937 Bush referred me to Thomas S. Knight, the Commercial Vice-President of the General Electric Company in New England. Knight, a member of the Cruising Club of America, had sailed Down-east all his life and was immediately attracted by a proposal to use recent developments in aerodynamics and other fields to harness the wind on a large scale. He offered to assist me in developing my ideas to the point where more accurate cost studies could be made. Under this arrangement, the aerodynamic outputs were computed from data secured in conferences with Dr. Theodor von Kármán, Director of the Guggenheim Aeronautical Laboratories at the California Institute of Technology. The wind regime was selected after conferences with Dr. Sverre Petterssen, Director of the Department of Meteorology of the Massachusetts Institute of Technology. Based on these assumptions, I considered and rejected many alternative designs and details, in the end arriving at the schematic layout that was finally built; and at a first approximation of the best dimensions for a large-scale wind-turbine of this design.

Professor John B. Wilbur, of the Massachusetts Institute of Technology

VANNEVAR BUSH

Electrical engineer, President of the Carnegie Institution of Washington, formerly Director of the Office of Scientific Research and Development, now Chairman of the Joint Research and Development Board of the Department of National Defense.

THOMAS S. KNIGHT

Commercial Vice President of General Electric Company in Boston since 1931.

(now head of the Department of Civil Engineering), also found his imagination stirred, and I retained him to carry out preliminary stress analyses. Having "frozen" the schematic layout, and Wilbur having determined the sizes of the members within limits, I retained Jackson and Moreland, Consulting Engineers of Boston, to make layout sketches and a brief report analyzing the economics of the project.

SVERRE PETTERSSEN

One time Director of the Weather Bureau at Bergen. One time Head of the Department of Meteorology, Massachusetts Institute of Technology. With British Air Ministry 1942–45, and now Chief of the Weather Forecasting Service in Norway.

Now, at this time there were two tactical problems that had to be solved before the wind-power project could become a going concern. First, it was necessary to find someone to put up the money for a full-scale test unit. Second, a manufacturer had to be found with the experience and prestige to qualify him for this project. Knight had granted me temporary office space in the Boston Offices of the General Electric Company and under his patronage I discussed the project, with Knight's consent, with various members of his staff including Alan Goodwin, his hydroelectric specialist. Out of the discussions with the latter came the suggestion that Goodwin see Walter Wyman, President of the New England Public Service Corporation, at Augusta, Maine, and clearly the one man in New England who had both the authority and the vision to push such a project as large-scale wind-power.

Goodwin, in connection with the various hydroelectric developments in which he was interested, had been dealing with Frank Mason, Chief Engineer of the New England Public Service Corporation. Goodwin knew that Mason was an enthusiastic believer in water-power and felt that he could be interested in wind-power if it could be demonstrated to him

JOHN B. WILBUR

Head of the Department of Civil and Sanitary Engineering, Massachusetts Institute of Technology, and Chief Engineer of the Wind-Turbine Project.

that wind-power would enhance the value of hydroelectric units.

Goodwin went to Augusta and talked to Mason, who recognized the great possibilities in a combination of wind and water. Mason looked out of his window, saw that Wyman's automobile was parked in its usual place, and so knew that Wyman was in his office. Mason also noted that the wind was one of those strong southwesterlies that frequently come during the hot summer spells of dry weather. Mason and Goodwin went to Wyman's office. Wyman, like Knight, was a Down-easter; he owned a farm on a windy ridge near enough the coast to come under the diurnal sea breeze in summer. The three looked out at the waving trees. It was a hot day in August and the water in the Utility Company's reservoirs was low. Goodwin said, "Mr. Wyman, just look at the way the wind is blowing those trees. There is a lot of force there and it is all going to waste. Man has used the wind for centuries to blow himself around the oceans but he has never harnessed it for power on a large scale. Wouldn't it be wonderful if you could use some of the power of the wind right now when it is hot and dry and windy and your water is running low?"

This made sense to Wyman, who also knew that Goodwin never talked without a purpose. He asked Goodwin what he had in mind and the upshot was that Goodwin returned to Boston able to tell his Chief, Knight, that Wyman wanted to discuss with Knight, at his earliest convenience, the purchase of a wind-power development for the New England Public Service System.

This conference had taken care of problem number one.

Problem number two was solved this way. During the preceding fifteen years Goodwin had been promoting hydroelectric developments in collaboration with representatives of water-wheel manufacturers. The most active New England representative was Howard Mayo of the S. Morgan Smith Company. Goodwin and Mayo realized that there was just so much water and, while they had not come to the end of their water in New England, they were interested in any means of enhancing the existing or the future water-power developments of New England. It seemed logical to them that the S. Morgan Smith Company, preeminent in the field of controllable-pitch hydraulic turbines, should take on the manufacture of a wind-power unit.

BEAUCHAMP E. SMITH

President, S. Morgan Smith Company.

Furthermore, Mayo knew that S. Morgan Smith Company was seeking to diversify its product. He took Goodwin to York, Pennsylvania, to see S. Fahs Smith, the President of the Company. The next week the two Vice-Presidents of the Company, Beauchamp Smith and Burwell Smith, came to Boston to see Knight and me.

There in Knight's office the project was born, in October, 1939, in the expectation of finding wind-power sites in New England where secondary but predictable power could be generated at the rate of 4400 kilowatt-hours per kilowatt per year, at a cost of $0.0025 per kilowatt-hour at the foot of the tower.

S. Morgan Smith Company decided that first they would redetermine the most economical dimensions and review my cost estimates. This work, described in Chapter IX, was carried out by a special staff of the Budd Company, in Philadelphia. The wind regime was re-specified by Petterssen. The aerodynamic outputs were recomputed by von Kármán. Some of the weights were estimated by Wilbur.

BURWELL B. SMITH

Vice-President and Treasurer, S. Morgan Smith Company.

Costs were obtained from vendors. The annual charges were set by Jackson and Moreland. The computations were carried out by Carl J. Wilcox, on loan to the S. Morgan Smith Company by the Budd Company. Variables considered included tower height, turbine diameter, turbine speed, generator rating, and generator speed.

The best design was found to be a 1500-kilowatt unit, driven at 600 revolutions per minute by a geared-up turbine 200 feet in diameter, turning at about 25 revolutions per minute, on a tower about 150 feet high. The energy cost at the switchboard at the foot of the tower was estimated to be $0.0016 per kilowatt hour.

Satisfied with this estimate, but aware that it was based on many assumptions which remained to be tested, the S. Morgan Smith Company agreed to engineer and build a test unit to my designs. General Electric had agreed to develop and furnish the electrical equipment at cost, and Walter Wyman had arranged that

Central Vermont Public Service Corporation, a subsidiary of the New England Public Service Company, should be the guinea pig, to provide the site and tie-in facilities, and should operate, and ultimately purchase, the test unit.

Fortunately both Albert A. Cree, President of the Central Vermont Public Service Corporation, and his chief engineer, Harold L. Durgin, were taken with the idea. Their system had no steam capacity. The hydro capacity was insufficient to carry peak load, and their excess

ALBERT A. CREE

President, Central Vermont Public Service Corporation. President, Windpower, Inc.

demand was met under a power-purchase contract with the Bellows Falls Hydro Electric Corporation. If the test unit of the Smith-Putnam Wind-Turbine proved successful, Wyman and Cree hoped to install a block of wind capacity to supplement their hydro capacity and reduce over-all power production and purchase costs.

In the late fall of 1939, S. Morgan Smith Company realized that they were not organized either to design the test unit of this new product or to fabricate it, because of the rapidly increasing backlog of hydraulic turbine orders, and that it would be necessary to sublet both the engineering and fabrication. Admittedly this was not a very satisfactory arrangement; but there was no alternative, and Beauchamp Smith asked me to find somebody to handle it. Several firms with outstanding reputations as successful designers of heavy precision rotating equipment were approached.

Alfred E. Gibson, President of the Wellman Engineering Company of Cleveland, Ohio, agreed to undertake the work on a cost plus contract. The Wellman Company are designers and manufacturers of large mining and material handling equipment.

At Beauchamp Smith's request, I became Project Manager. It was my responsibility to organize the project and

HAROLD DURGIN

Vice-President and Chief Engineer, Central Vermont Public Service Corporation.

conduct its external relations, in accordance with the general policies laid down by Smith. He encouraged gathering together as consultants the leading authorities in the various related fields, to make possible a coordinated attack upon those problems in meteorology, ecology, aerodynamics, vibration analysis, electrical engineering, and the economics of power generation, which in 1939 remained to be solved before a large wind-turbine could be successfully and profitably operated on the lines of a utility system.

George A. Jessop, Chief Engineer of the S. Morgan Smith Company, was active in general supervision of the project, and contributed to the solution of many of the novel problems encountered.

Dr. John B. Wilbur, who later became Chief Engineer of the Project, heading the engineering and development and reporting to Jessop, was retained as structural consultant. The forces coming into the structure from the turbine were estimated from data supplied by von Kármán, who was retained as a general consultant, specifically responsible for the aerodynamic design.

GEORGE A. JESSOP

Chief Engineer, S. Morgan Smith Company. Member A.S.C.E. and A.I.E.E.

It was under von Kármán's supervision that Dr. Elliott G. Reid, of Stanford University, took charge of wind-tunnel tests of models of the turbine blades. The models had been constructed, some in California, some by the S. Morgan

Smith Company in York, Pennsylvania, but all under the supervision of Mr. John Haines, now Director of Research of the Aeroproducts Division of General Motors. Haines served the project as a mechanical engineer in charge of the problems of pitch control and coning.

Professor J. S. Newell, of the Aeronautical Engineering Department of Massa-

ROBERT F. GRIGGS

Of George Washington University, Chairman of the Division of Biology and Agriculture, National Research Council.

chusetts Institute of Technology, was consulted concerning the structural analysis of the turbine blades, whose Cor-Ten spars were designed by the American Bridge Company. The stainless-steel skin and ribs of the blades were designed by the Budd Company, noted for their pioneer work with stainless-steel trains. Budd fabricated the stainless-steel parts and assembled them around the spars received from the American Bridge Company.

The mechanical design, from the blade roots to the tower cap, was carried out by the Wellman Engineering Company under the direction of R. W. Valls, while the tower was designed and erected by the American Bridge Company.

Vibration analysis of the entire structure was placed in charge of Dr. J. P. den Hartog, at that time of Harvard University, now at Massachusetts Institute of Technology.

According to the agreement, the electrical equipment was supplied by the General Electric Company.

The meteorological program was directed by Petterssen, in collaboration with Dr. Hurd Willett, also of Massachusetts Institute of Technology; Dr. C. F. Brooks, Director of Harvard's Blue Hill Observatory; and Dr. K. O. Lange, also of Harvard.

Wind-tunnel tests of models of mountains were carried out by Dr. Th. Troller of the Guggenheim Aeronautical Institute, at Akron, Ohio, under the direction of von Kármán.

HURD CURTIS WILLETT

Professor of Meteorology at the Massachusetts Institute of Technology.

The ecological studies were carried out by Dr. Robert F. Griggs of George Washington University.

Jackson and Moreland served as application engineers, and from time to time carried out various studies of the economics of wind-power, in collaboration with the Central Vermont Public Service Corporation.

The Smith-Putnam Wind-Turbine Project

Beauchamp Smith had insisted on gathering together the leading men in each field; without a doubt he had succeeded. Each in turn came under the peculiar spell of the project, which roused the enthusiasm of all of us. The tone set by the Smiths was a little unusual. The realism with which they approached the project was garnished with dash—even gaiety. This spirit soon infected all who were engaged on the project, and produced an esprit de corps which went far toward counterbalancing the disadvantages that all were not housed under one roof, and indeed were not all working on the project full time. The project was administered by means of frequent gatherings in various cities, interspersed with telephone conferences. It was no easy task for Dr. Wilbur to weld together into one functioning whole all of these components, when the key personnel were scattered

CHARLES F. BROOKS

Professor of Meteorology, Harvard University, and Director, Blue Hill Observatory.

from coast to coast, in many cases with their attentions partly diverted to other consulting problems.

One of the critical aspects of the project was the control of the machine in all phases of operation. It was necessary to select items from among the mechanical, hydraulic, and electrical control apparatus already in use in other fields, principally in connection with hydroelectric units, and to adapt the most suitable to our needs. The selection, adaptation, and interconnection of the various control elements required careful consideration. Various alternative methods of control were devised before the final control details were determined. To head up and centralize all control problems, Grant H. Voaden, of the S. Morgan Smith Company, was appointed Chief Test Engineer.

KARL O. LANGE

One time Research Associate at Massachusetts Institute of Technology, and Research Meteorologist at the Blue Hill Meteorological Observatory. Now Professor of Physics, University of Kentucky.

Voaden, who had had considerable experience in this class of work in the hydraulic turbine field, worked in collaboration with Irl Martin and Walter Thorell of the Woodward Governor Company and George Jump and Herman Bany of the General Electric Company.

In May, 1940, we began ordering steel forgings. In June, Grandpa's Knob was selected as the test site, because, having an elevation of only 2000 feet, it was, we

9

hoped, not high enough to encounter destructive ice storms. The 2-mile road connecting the site with the Vermont highway network was built that summer, under the direction of the Central Vermont Public Service Corporation, and the tower was erected by the American Bridge Company.

Stanton D. Dornbirer of the Erection Department of S. Morgan Smith Company was placed in charge of the field erection of the wind-turbine.

Erection of equipment aloft continued throughout the winter, as weather permitted. Low temperatures and high winds made rough work of handling heavy steel, but we had only one accident. On a particularly bitter, windy sub-zero day, the 40-ton pintle girder (Fig. 1) with the 24-inch main shaft and its two main bearings in place, was being trucked to the summit, on a heavy trailer drawn by the truck unit and two Caterpillar tractors in tandem. At the hairpin turn below the summit the girder rolled off the trailer-bed and turned upside-down into a deep snowbank, the 48-inch main-bearing housings having found the only opening between rocks! After recovery, inspection showed no damage to the bearings or shaft, and only minor damage to some of the plates of the girder.

Erection was completed in August, 1941, when the blades were put in place.

To facilitate carrying out the test program a field office was set up in Rutland in July, 1941. Activities of this office were under the direction of Dr. Wilbur and his assistant, G. H. Voaden. The office was staffed with three test engineers and a secretarial force.

In June, 1940, I had gone to Washington, ultimately to join Dr. Bush and the Office of Scientific Research and Development, and was not again active in the wind-turbine project until the spring of 1945.

When the turbine was assembled, William Bagley, of the General Electric Company, in collaboration with Voaden, checked the controls and carried out a number of test runs at no load, beginning at low speed and progressively altering the pitch until, after a few weeks, the turbine was run at the full speed of 28.7 revolutions per minute, without load.

Finally, on the night of Sunday, October 19, 1941, in the presence of the top management of the Central Vermont Public Service Corporation and many of the Staff of the S. Morgan Smith Company, and with the Smiths listening in by long-distance telephone, Bagley, having completed his adjustments and made his final inspections aloft, phased-in the unit to the lines of the utility company, in a gusty 25-mile wind from the northeast.

There was no difficulty. Operation was smooth. Regulation was good. After 20 minutes at "speed-no-load," the blade pitch was adjusted until output reached 700 kilowatts. For the first time anywhere, power from the wind was being fed synchronously to the high-line of a utility system.

Then began months of gathering experience, and of making adjustments leading to smoother operation. A comprehensive test program was put on a routine basis, in charge of Grant Voaden, reporting to Wilbur.

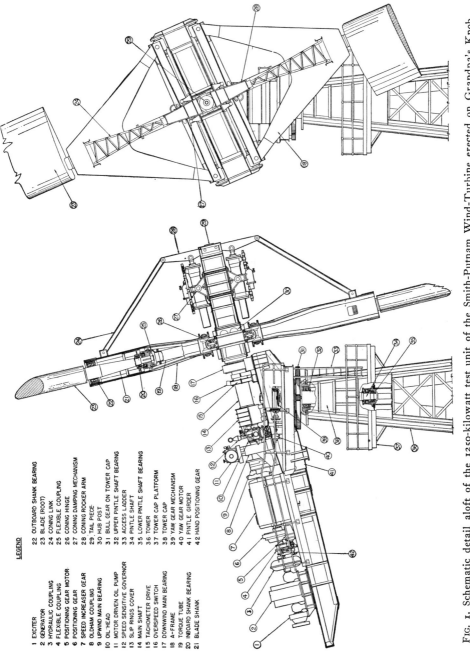

LEGEND

1 EXCITER
2 GENERATOR
3 HYDRAULIC COUPLING
4 FLEXIBLE COUPLING
5 POSITIONING GEAR MOTOR
6 POSITIONING GEAR
7 SPEED INCREASER GEAR
8 OLDHAM COUPLING
9 UPWIND MAIN BEARING
10 OIL HEAD
11 MOTOR DRIVEN OIL PUMP
12 SPEED SENSITIVE GOVERNOR
13 SLIP RINGS COVER
14 MAIN SHAFT
15 TACHOMETER DRIVE
16 OVERSPEED SWITCH
17 DOWNWIND MAIN BEARING
18 A-FRAME
19 TORQUE TUBE
20 INBOARD SHANK BEARING
21 BLADE SHANK
22 OUTBOARD SHANK BEARING
23 BLADE (ROOT)
24 CONING LINK
25 FLEXIBLE COUPLING
26 CONING HINGE
27 CONING DAMPING MECHANISM
28 CONING ROCKER ARM
29 TAIL PIECE
30 HUB POST
31 BULL GEAR ON TOWER CAP
32 UPPER PINTLE SHAFT BEARING
33 ACCESS LADDER
34 PINTLE SHAFT
35 LOWER PINTLE SHAFT BEARING
36 TOWER
37 TOWER CAP PLATFORM
38 TOWER CAP
39 YAW GEAR MECHANISM
40 YAW GEAR MOTOR
41 PINTLE GIRDER
42 HAND POSITIONING GEAR

FIG. 1. Schematic detail aloft of the 1250-kilowatt test unit of the Smith-Putnam Wind-Turbine erected on Grandpa's Knob.

In the course of this program the test unit was operated in winds of 70 miles an hour, generating as much as 1500 kilowatts, and it was exposed, while not operating, to winds of 115 miles an hour. Regulation was found to be satisfactory. Only light icing was encountered during operation. Under these conditions the stainless-steel blades shed ice well. Under certain conditions there was some vibration, which was satisfactorily reduced by adjustments to the coning-damping system and the yaw-damping system, and largely eliminated in the plans for redesign described in Chapter X.

The first major mishap occurred when, in February, 1943, the 24-inch downwind main bearing failed, for reasons which are still obscure, although apparently not related to anything peculiar to wind-turbines. Because of the war it took more than twenty-four months to obtain and install a new one. But these months were not wholly lost. The skeleton staff were busy digesting observational data, and designing simplifications and improvements, many of which were carried out when the new bearing went on.

One of the most important studies made in this period was a review of loadings, by Wilcox, in collaboration with Holley, under the direction of Wilbur. They concluded that some of the loadings had been over-estimated and some under-estimated. As a net result, they felt they could now detail the design of a large wind-turbine, with conventional factors of safety throughout, and which would weigh less per kilowatt than the test unit, and cost less per pound. But, of immediate concern, they concluded that the stresses in the root sections of the blades of the test unit were higher than had been realized. Accordingly, Wilbur, early in 1945, recommended that the unit should not be operated after it had completed its test purposes, and he hoped for an early completion of this program.

With the new bearing in place, the Smith-Putnam Wind-Turbine went back on the line on March 3, 1945. Wind permitting, it was continuously operated as a routine generating station on the lines of the Central Vermont Public Service Corporation for twenty-three days. Operation was satisfactory. There was no trouble.

At 5 A.M. on March 26, 1945, my telephone started ringing. It wouldn't stop. It was Wilbur, "Put, we've had an accident. It could be worse. We've lost a blade, but no one is hurt, and the structure is still standing." I got Beauchamp Smith out of bed in York and reported the accident.

This is what had happened. In 1942, after a period of operation, the blade skins had begun to break near the roots. It was necessary to reinforce the blades at the root section. But the blade shank spar connection was already a source of anxiety since it was known to be a very highly stressed member, and stiffening the blade skin near the root would throw still greater stresses into the spar section at its weakest point. It was a Hobson's choice, and the decision was to carry out the modifications in the field. Otherwise it would have been necessary to discontinue the field test program until such time as postwar blades could be fabricated, delivered, and erected; and that choice, as it turned out, would have meant a delay of some four years. Subsequent study has shown that the field welding associated with these repairs to the blades had, as had been feared, in effect cut deep notches in the blade spars in the very zone where they were known to be most highly stressed. Stress concentrations, resulting from these inadvertent notches, had caused the spars to crack progressively. But the cracks had occurred just behind a bulkhead which made inspection impossible. At 3:10 on the morning of March 26 the turbine was operating in a smooth, steady, southwest wind of about 25 miles per hour. The spars now had hidden cracks across more than 90 per cent of the cross-sectional area. The tension of centrifugal force, amounting to several hundred thousand pounds, was being withstood by only a few square inches of Cor-Ten steel.

Harold Perry, who had been the erection foreman, and was a powerful man, was on duty aloft. Suddenly he found himself on his face on the floor, jammed against one wall of the control room. He got to his knees and was straightening up to start for the control panel, when he was again thrown to the floor. He collected himself,

FIG. 2. The blade that failed.

got off the floor, hurled his solid 225 pounds over the rotating 24-inch main shaft, reached the controls, and brought the unit to a full stop in about 10 seconds by rapidly feathering what was found to be the remaining blade of the turbine. He estimates that it took him about 5 seconds to get to the controls after the first shock.

One of the 8-ton blades had let go when in about the 7 o'clock * position, and had been tossed 750 feet, where it landed on its tip (Fig. 2).

* Subsequently calculated by von Kármán.

Following the failure, Beauchamp Smith asked me to review the entire project, and to estimate the future of large-scale wind-power.

These studies were carried out by Carl J. Wilcox and Stanton Dornbirer, in collaboration with Jackson and Moreland, and in consultation with George A. Jessop, John B. Wilbur, and Myle J. Holley.

If the test unit, modified and "cleaned-up," but not redesigned for production, had been put into limited production in the fall of 1945, a block of 6 could have been installed on Lincoln Ridge (4000 feet), in Vermont, at a cost of about $190 per kilowatt. However, the worth of this block of 9000 kilowatts to the Central Vermont Public Service Corporation, as estimated by Jackson and Moreland, but not confirmed by the Corporation, was about $125 per kilowatt.

Several means for bridging the gap between $190 and $125 were suggested and are discussed in Chapter XII. It would cost several hundred thousand dollars to test these suggestions, however, and S. Morgan Smith Company found that the scientific research and development already carried out had cost over a million and a quarter dollars—far more than had been originally estimated by me or them. Being a small company, the Directors felt it would be imprudent to lay out further substantial funds, with no reasonable guarantee of economic success in the end. Accordingly, in November, 1945, they decided, with the concurrence of Wilbur and Jessop, to abandon the project.

In the wake of this decision, Beauchamp E. Smith and Burwell B. Smith asked me to write this account of how a small company undertook a risky and expensive exploration beyond the frontiers of knowledge—of how they succeeded technically but could not afford to find out if they had succeeded economically. The Smiths wish to make available to the world the knowledge that has been gained, in the hope that someone else, perhaps in windy Scotland, or Ireland, or New Zealand, or southern Chile, or any place where winds are strong and fuels scarce, may see fit to continue from where, regretfully, they have been forced to halt.

Chapter II

HOW DOES WIND BEHAVE? OUR ASSUMPTIONS OF 1939

Introduction

The density of the air and the average velocity of the wind are low compared with other commercial propelling fluids. The kinetic energy of a unit volume of moving air is less than that of falling water or high pressure steam, for example. Accordingly, the projected area of turbine blades to be driven by the wind must be greater, per kilowatt of output, than in the case of a hydraulic turbine or a steam turbine. The structure to support this great turbine, when designed to withstand ice and wind storms, runs to great weight, and therefore to high installed cost, per kilowatt. For example, the test unit of the Smith-Putnam Wind-Turbine weighed about 500 pounds per kilowatt.

Thus the low density and velocity of the wind are handicaps which stand in the way of its use as an economical prime mover. It is not feasible in this case to improve the density by supercharging, but nature can be made to assist in the matter of the velocity. There are mountain ridges which act like airfoils, and which speed up the normal flow of the wind. Other types of mountain retard the flow.

An intimate knowledge of the habits of wind-flow will permit one to select a site for a turbine where the free-air velocity has been speeded up 20 per cent or more. The power in the wind varies as the cube of the velocity. For example, the power in a 20-mile wind is 8 times greater than the power in a 10-mile wind. Obviously, it is important to find a site which speeds up the wind by a factor of, say, 1.20, rather than one which may retard it.

The compression of the streamlines over a good airfoil tends to damp out turbulence and provides a further incentive to learn in accurate detail the habits of wind-flow, particularly in mountainous country.

Such accurate and detailed knowledge is necessary if the designers are to select a site where the handicaps inherent in this slow-moving and tenuous propelling medium are reduced so far as may be, and if they are to develop a design which can work on advantageous terms in economic harness with water-power and steam stations.

It will be seen shortly how far we were in 1939 from possessing a knowledge of the wind adequate for site selection, or for the economical design and operation of a large wind-turbine.

Major Circulations of the Wind

The sun is the ultimate source of the energy in the wind. The atmosphere enveloping the earth is a rotating, regenerative thermal engine, stoked by radiant

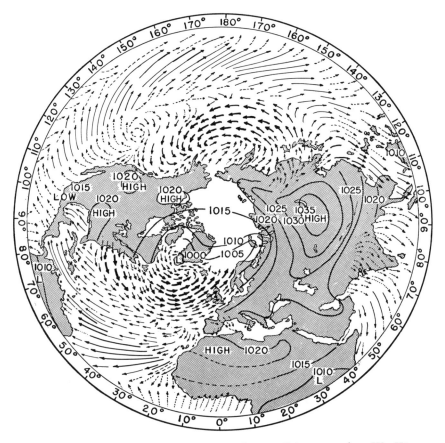

FIG. 3. Prevailing winds over the oceans, January-February, after W. Köppen. Width of arrow indicates strength of wind.

> ---→ Less than 10 miles an hour.
> ——→ From 10 to 15 miles an hour.
> ——→ From 15 to 30 miles an hour.
> ——➤ Over 30 miles an hour.

Length of arrow indicates steadiness of wind.

energy from the sun. The complex dynamics of this system result in the types of major circulation shown in Figs. 3 through 6, comprising the doldrums at the equator, the northeast trade winds in the tropics, the calms of the horse latitudes, and the prevailing westerlies of the "roaring forties." This information, supplemented by 50 years of U. S. Weather Bureau records from ships at sea, has been analyzed by Petterssen (Ref. 1-B), as set out in Table I and summarized in Fig. 7, the velocities being converted to potential power outputs at various oceanic sites.

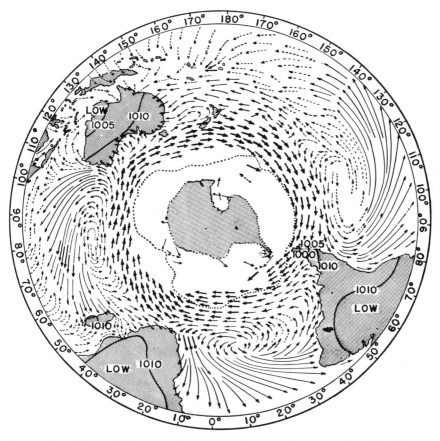

Fig. 4. Prevailing winds over the oceans, January-February, after W. Köppen. Width of arrow indicates strength of wind.

----→ Less than 10 miles an hour
——→ From 10 to 15 miles an hour.
———→ From 15 to 30 miles an hour.
━━━▶ Over 30 miles an hour.

Length of arrow indicates steadiness of wind.

Where land masses are absent, as in the Antarctic seas, the wind circulation is strong and fairly uniform. But, where land masses are prevalent, the atmospheric envelope is called upon to act as a heat exchanger between the thermally nearly static oceanic areas and the alternately winter-chilled and summer-heated land areas, producing a more erratic general circulation, which is further and profoundly influenced by local topography.

Various continental regions are known to be windy. In general, they lie between the latitudes of 25 degrees and 60 degrees and are mountainous. These areas

FIG. 5. Prevailing winds over the oceans, July-August, after W. Köppen. Width of arrow indicates strength of wind.

> ---➤   Less than 10 miles an hour
> ——➤   From 10 to 15 miles an hour.
> ——➤   From 15 to 30 miles an hour.
> ━━➤   Over 30 miles an hour.

Length of arrow indicates steadiness of wind.

are listed in Table II. Information is generally lacking on which to base reliable estimates of annual output in these regions.

Limits of Our Knowledge of Available Wind-Power, in 1939

This gross survey of oceanic sites, with some indications of the windiness of certain coast lines, represents about the limit to which in 1939 we could develop a detailed knowledge of wind-power.

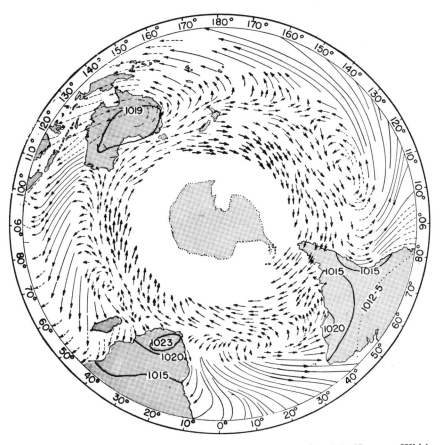

FIG. 6. Prevailing winds over the oceans, July-August, after W. Köppen. Width of arrow indicates strength of wind.

- ····>   Less than 10 miles an hour
- ——>   From 10 to 15 miles an hour.
- ———>   From 15 to 30 miles an hour.
- ———>   Over 30 miles an hour.

Length of arrow indicates steadiness of wind.

Our Problem in 1939

In 1939 our problem was to answer two practical questions:

Question 1. Which is the best test site in Vermont?

19

TABLE I. WINDY ISLANDS AND COASTS
Classified by Annual Output in Kilowatt-Hours per Kilowatt *

Region	Output in Kwh./Kw.								
	Over 6000	5000 to 6000	4500 to 5000	4000 to 4500	3500 to 4000	3000 to 3500	2500 to 3000	2000 to 2500	Less Than 2000
North Atlantic Ocean		Iceland	Faeroes Ireland Scotland Shetlands Norway	Azores Bermuda	Devonshire Wales	Cape Verde Island Martinique Porto Rico	Trinidad Aruba St. Martin	Cape Hatteras Madeira Nantucket Newfoundland Nova Scotia Santo Domingo	Bahamas Cuba Haiti Jamaica St. Martins Colon
South Atlantic	Sandwich Is. South Georgia	Falkland Is. Gough	Tristan da Cunha			Ascension St. Helena	Fern. de Noronha So. Africa	St. Paul	
North Pacific Ocean			Aleutians	Kamchatka Kuriles	Japanase Island	Formosa Palmyra		Hawaii Philippines	Guam
South Pacific Ocean	Auckland Campbell Macquarii	Bounty	Tasmania Ware-kauri	Chile New Zealand	Australia Lord Howe	Norfolk Ryukas	Kirmadec New Caledonia	Easter Fiji Gilberts Juan Fernández Luisiades Malden Marguesas Marshalls New Hebrides Pitcairn St. Felix Sala-y-Gomez Tonga	Samoa Society Is. Santa Cruz
Indian Ocean		St. Paul		Mauritius	Cocos		Madagascar	Ceylon Socotra	Andamans Seychelles Tachago

* Based on a 1250-kilowatt turbine, 175 feet in diameter. If a turbine operated continuously at 100 per cent of capacity, the annual output would be 8760 kilowatt-hours per kilowatt (24 hours per day for 365 days).

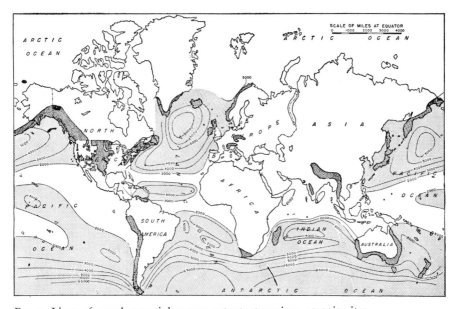

Fig. 7. Lines of equal potential power output at various oceanic sites.

Developed by Sverre Petterssen, largely based on fifty years of United States Weather Bureau Records from ships at sea.

Outputs are stated in kilowatt-hours per kilowatt per year from a 1250-kilowatt turbine, 175 feet in diameter. The annual output of a turbine operated continuously at 100 per cent capacity would be 8760 kilowatt-hours per kilowatt (24 hours per day for 365 days).

Table II. Windy Continental Regions

North America
    Labrador and Maritime Provinces of
      Canada
    New England States
    Great Lakes Area
    Great Plains Area
    Pacific Coast
    Alaska

South America
    Andes
    Venezuela

Europe
    Alps of Central Europe
    Apennines of Italy
    Guadarramas of Spain
    Pyrenees of Spain and France
    Carpathians of Poland

Asia
    Urals of Russia
    Himalayas of India and Tibet
    Deccan of India
    Kamchatka Peninsula of Russia

Antarctica

Question 2. Which is the best site in Vermont for ultimate exploitation to about 20,000 kilowatts of installed capacity?

The answers involved a third question:

Question 3. What are the meteorological factors which influence wind-turbine design and site selection, and how can they be evaluated?

It has sometimes been said that "wind is wind," blowing with more or less uniform strength, and people have wondered why it is necessary to take pains about the precise location of a wind-turbine site. Even if the surface of the ground were billiard-table flat, wind would not be wind in this sense, because of the frictional retardation, which is especially great within 50 feet of the surface. But our test unit was to be erected in the mountainous state of Vermont, and we knew just enough to realize that in such terrain wind is not wind, and that there was much about the habit of its flow to be determined before a large-scale wind-turbine could be efficiently designed or operated. At that time we had no fore-warning of how complex the problem was to become, and of how elusive some of the answers were going to be.

Our Basic Assumptions of 1939

To launch the project at all, to decide upon a general design, and to estimate operating profits, it had been necessary to ask the leading meteorologists and aerodynamists to make first approximations to the answers to Question 3, based on the state of knowledge in 1939.

In response there was developed a set of working assumptions, the basis on which the Project was launched (Ref. 2-B). These concerned:

1. The free-air wind velocity at mountain-top height in Vermont.
2. The effect of the geometry of a mountain upon the retardation or speed-up of wind-flow over its summit.
3. Prevailing wind directions in the western foothills of the Green Mountains.
4. Influence of the structure * of the wind on design.
5. Influence of the structure of the wind on estimates of output.
6. Influence of atmospheric density on estimates of output.
7. Influence of estimates of icing on design and on site selection.

1. *The Free-air Wind Velocity at Mountain-top Height, in Vermont.*

It is not possible to estimate retardation or speed-up of air-flow by a mountain top unless one knows, or can assume, a value for the undisturbed free-air velocity at the elevation of the mountain top. The difficulty is that it may be impossible to find a truly undisturbed free-air velocity at any point up-wind of the ridge, at the elevation of the ridge, since the free flow will be affected by the broken country lying underneath it.

---

* As will be described later, wind, especially near the ground, is turbulent and gusty, and changes rapidly in direction and in velocity. This departure from homogeneous flow is collectively referred to as "the structure of the wind."

Increase of wind velocity with height above the ground. Even over such flat country as the level, baked-clay surface of the Great Australian Desert, the flow of the free-air is retarded by the friction of the ground. Since the frictional force decreases with elevation, the wind velocity will increase as one ascends through the atmosphere.

The height at which the wind blows unimpeded by ground effects—that is, the gradient wind level. The rate at which velocity increases with height is greatest near the ground, decreasing as one leaves the ground. It has been common practice in the literature of meteorology to assume that over flat country that rate of change of velocity with elevation that is due to decreasing frictional effects will normally become negligible at a height of about 1000 meters above the ground. For practical purposes this height is known as the gradient wind level; but, over rough country, the thickness of this friction layer is normally greater. When the wind velocity is high and the temperature decreases strongly with altitude, there is much turbulent mixing, and therefore good mechanical interlocking between the ground and the air mass, and the influence of surface friction will extend upward indefinitely. In these circumstances the concept of a definite height (gradient wind level), at which the wind blows unimpeded by ground effects, vanishes.

Estimating the vertical distribution of velocity between the gradient wind level and the ground. The wind velocity at the gradient wind level can be computed from the isobars * on the daily surface weather map. Knowing the wind velocity so computed at gradient wind level, and making some arbitrary allowance for the frictional effect due to a particular terrain, one can estimate the distribution of velocity between the ground and the gradient wind level.

Whatever the height of the gradient wind level may be, it was assumed that the "free-air" velocity just up-wind of a mountain top varied somehow with the elevation above sea level, and also with the relative elevation above the surrounding country. The true situation was found to be far more complex.

General lack of suitable wind-velocity data. Information was not available in 1939 with which to prepare a map showing the strength and constancy of winds over the summits of potential wind-power sites in New England; nor was it possible to estimate these factors by extrapolation from U. S. Weather Bureau anemometer readings, pilot-balloon observations, or pressure maps, except in the roughest sort of first approximation.

Lacking an actual wind-energy survey, various interested persons had from time to time attempted to work up the U. S. Weather Bureau chart of surface wind velocities (Fig. 8). This chart is an accurate summary of the mean annual velocities and directions, observed at the various U. S. Weather Bureau Stations (Ref. 1-A). But isovents † passing through these points are meaningless guides to wind-power, because U. S. Weather Bureau stations are rarely located on poten-

* Isobars are lines of equal barometric pressure.
† Lines of equal mean annual velocity.

23

tial wind-power sites, which are exposed high ridges lying across the prevailing wind. Weather Bureau stations are usually located either on buildings in cities or at airports. Airports are not located on mountain tops. The velocity readings at stations in cities are influenced by the turbulence and frictional drag of buildings. Readings at many stations are affected by the erection of new buildings, the growth of the cities, and the incorrect location of the anemometer in the wind

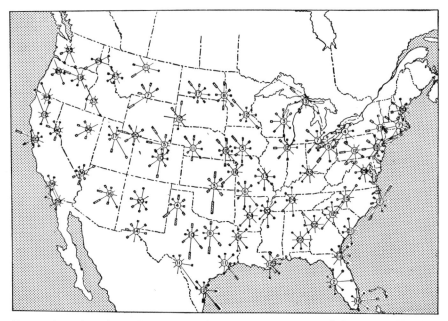

FIG. 8. Annual surface wind roses. United States Department of Commerce, Weather Bureau.

The numbers in the circles refer to the percentage of calms (0–3 miles per hour). The length of the spokes indicates the steadiness of the wind.

shadow of sloping roofs or of other instruments. The velocity readings, therefore, are valid only for extremely localized conditions, and it is not possible, for example, to estimate the velocities at 140 feet over Mt. Abraham, in the Green Mountains, by extrapolation from the velocities recorded by the Weather Bureau in the city of Albany, 110 miles away.

The wind velocity data available. The extent of our knowledge concerning the particular wind regime prevailing over the mountains of New England was limited to anemometry from two mountain stations operated under the direction of C. F. Brooks—Mount Washington Observatory (6288 feet) in northern New Hampshire, and Blue Hill Observatory (635 feet), 10 miles south of Boston; and to upper air data from the only two pilot-balloon stations in New England—

Boston and Burlington. These sets of data, combined with certain assumptions, had been used to develop a point of departure, as follows.

Our first assumptions about wind in the mountains of New England. We established the respective wind velocities at anemometer height above the summits

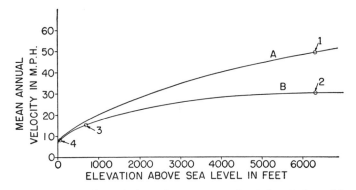

Fig. 9. Assumed variation of mean annual wind velocity with variation in elevation above sea level, in New England.

Curve A represents velocities as accelerated by mountain-tops, the example being the value reported in 1935 for Mt. Washington (6300 feet), of about 49 miles an hour, since found to be high by about 5 miles an hour.

Curve B represents free-air velocities.

    Point 1   Long-term mean velocity as reported in 1935 by Mt. Washington Observatory.

    Point 2   Free-air mean velocity at elevation of Mt. Washington, computed by Brooks.

    Point 3   Long-term mean velocity at Blue Hill Observatory, reported by Brooks in 1935.

    Point 4   Long-term mean velocity at East Boston Airport (sea level), reported by Brooks in 1935.

of Mount Washington and Blue Hill by taking the average velocity over a 5-year period of anemometry at each station. We estimated the free-air velocity at the same elevations from the data of Boston pilot-balloon observations and pressure maps, and, to establish an end point, by anemometry records of East Boston Airport (35 feet). This work was carried out by C. F. Brooks. In Fig. 9 the mean velocities thus obtained are plotted against elevation above sea level.

Brooks concluded that the speed-up factor at Blue Hill was close to unity (1.0), that is, that the wind passing over Blue Hill at anemometer height had been neither speeded up nor retarded. Accordingly, a smooth curve was drawn connecting the three "fixes" of the free-air velocity. This (lower) curve of Fig. 9 represents the assumed variation of the free-air velocity with elevation above sea level in New England. Mountain-top velocities falling on this lower curve would correspond to speed-up factors of unity (no speed-up or retardation). Points falling below this curve would characterize the tops of mountains whose masses had retarded the free-air, and where the speed-up factors were therefore less than unity. Points falling above this line would characterize mountain tops which had speeded up the wind (speed-up factor greater than unity).

Now, Brooks had found on Mt. Washington that the speed-up factor increased as the free-air velocity increased. So we assumed that, for aerodynamically identical mountain crests, the speed-up factor would vary somehow with velocity and, accordingly, a crude limiting envelope was suggested by drawing the smooth upper curve, which tended to converge with the lower at the lower velocities. Our assumption was that, in any 100 square miles of the Green Mountains, we would not fail to find some high ridge, the velocity on whose summit would fall between these two curves. Specifically, we estimated that we could find such ridges at 4000 feet with speed-up factors of at least 1.20. This we have probably done (Lincoln Ridge).

The uncertainty about the Burlington data. The analysis of the Boston and Burlington pilot-balloon observations, that led to the free-air velocities entered in Fig. 9, presents a vivid picture of the difficulties and uncertainties encountered in using such observations for our purpose.

The accumulated data from pilot-balloon observations over Boston showed the expected continual increase of wind velocity with elevation. The pilot-balloon observations over Buffalo, which were used as a check, showed essentially the same characteristic. We realized that pilot-balloon measurements are selective measurements. Before the advent of radar, determination of the wind by pilot-balloons eliminated all data with cases of weather in which the balloons disappeared in fog or clouds. Such weather, more often than not, is windy. Thus, windy data are eliminated from the records, more and more as the elevation increases. It was, therefore, felt that the upper-air wind velocities indicated by pilot-balloon material were in general lower than the actual and could be used safely in our velocity estimates.

The Burlington pilot-balloon data, however, indicated a distribution of velocities with elevation quite different from that found over Boston. Instead of a steady increase of velocity with elevation, the Burlington data indicated that the wind velocity increased up to about 1500 feet and then remained almost constant up to about 3000 feet (Fig. 10). By comparison with Boston it was obvious that the Burlington pilot-balloons represented a particular local condition. But how should it be interpreted? Was it characteristic of velocities over foothills up-wind of mountainous terrain or was the mountainous terrain responsible for a systematic error of measurement?

Now an inherent weakness of the standard pilot-balloon method * is that the velocity data become unreliable whenever the balloon rises through air with strong

* Following the balloons with two theodolites, one at each end of a base line.

vertical motion, in the sense that the indicated wind velocity is too low when the balloon rises faster than usual. There obviously exists a field of up-currents to the windward of the Green Mountains. Pilot balloons released at Burlington into the prevailing westerly wind would float directly into this up-current field. In this manner the wind velocities indicated by Burlington pilot balloons would become too low after a few minutes, that is, after the balloon had risen to a few thousand feet.

The interpretation on which the project was based. The consensus of the consulting meteorologists was that this systematic error in the pilot-balloon evaluations at Burlington was responsible for part, if not all, of the peculiar characteristics of the Burlington wind curve. It was, therefore, decided to base the free-air velocities of Fig. 9 on the Boston pilot-balloon data alone.

The curves of Fig. 9 were used as the working hypothesis by which we estimated that the mean annual velocity at the 2000-foot test site, Grandpa's Knob, would be 24 miles an hour. It proved to be about 17 miles an hour.

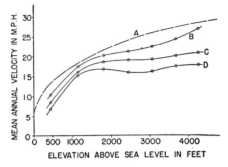

Fig. 10. Actual variation of mean annual wind velocity with variation in elevation above sea level as recorded by United States Weather Bureau pilot-balloon data at Burlington, Vermont.

Curve A represents the original assumption (Curve B of Fig. 9).

Curve B shows the actual distribution in winter afternoons.

Curve C shows the mean distribution for the year.

Curve D shows the actual distribution for summer mornings.

It was not until 1945 that Lange, in view of the unexpectedly low output at Grandpa's Knob, suggested and carried out a new and detailed analysis of the Burlington pilot-balloon material, which showed that the Green Mountain Range actually does exert its influence on the wind far out to windward, over the foothills. Lange's study is summarized in Chapter IV, and explains in part the failure to obtain the originally predicted wind velocities on Grandpa's Knob—a failure which resulted in an output there amounting to only 30 per cent of what we had predicted, based on Fig. 9.

2. *The Effect of the Geometry of a Mountain upon the Retardation or Speed-up of the Wind-Flow over Its Summit.*

Prior work, especially in gliding and in certain wind-tunnel programs, indicated approaches to this problem.

It was assumed that a north-south * ridge with a suitable profile would speed up the flow of free-air over its summit, by a factor of at least 1.20. (The factor at Mount Washington was thought to be in excess of 1.50.)

* We assumed a prevailing west wind in Vermont.

Consider a ridge of uniform cross-section: If the ridge does not lie parallel to the wind, it will act as a baffle and tend to deflect the wind.

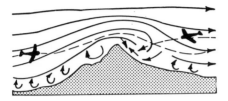

FIG. 11. Assumed behavior of wind flowing over a ridge. From Petterssen's "Introduction to Meteorology," Fig. 78, page 118.

As the angle increases, the effect on the wind direction increases. When the angle becomes 90 degrees and the ridge lies athwart the wind, like a dam, some of the wind will flow over the ridge, but some will stream around either end. The distribution will depend in part on how much more work is required to deflect a parcel of wind vertically than to deflect it horizontally. And this depends on the thermodynamic characteristics of the air

mass. This reasoning led us to suppose that such a ridge as the Green Mountains (160 miles long from north to south) would cause significant differences in the prevailing wind direction at the northern and southern ends. As stated in Chap-

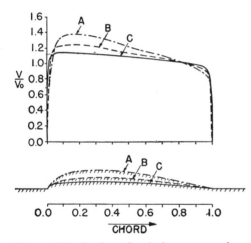

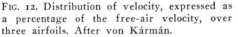

FIG. 12. Distribution of velocity, expressed as a percentage of the free-air velocity, over three airfoils. After von Kármán.

The maximum thicknesses of the three airfoils expressed as percentages of the respective chord lengths are:

Airfoil	Thickness
A	12.0%
B	7.5%
C	4.5%

ter IV, the effects actually found were larger than anticipated, and of great importance in site selection.

It was assumed that wind flowing over a ridge would tend to behave as shown schematically in Fig. 11. But how critical for a wind-turbine is the zone of rupture on the lee side? No answer was available. This was something we had to determine.

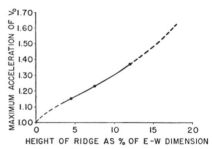

FIG. 13. Variation in the maximum acceleration of the wind velocity over a ridge, with variation in the profile of the ridge.

Von Kármán showed that, on ridges with steep western faces (Fig. 12), the speed-up of the wind over the summit would be greater than when the ridge is high compared to the east-west width of the base. This is summarized in Table III and Fig. 13.

He thought that the maximum value of the accelerated velocity would be found just at the crest of the ridge, but it was not certain that the turbulent flow found

TABLE III. EFFECT OF THE GEOMETRY OF A RIDGE UPON THE ACCELERATION
AND THE FREE-AIR VELOCITY

Height of ridge as per cent of east-west width of base of ridge	$\dfrac{V_{\text{accelerated}}}{V_{\text{free-air}}}$
4.5%	1.15
7.5%	1.23
12.0%	1.37
17.5%	1.6

in nature over wooded and broken ridges would conform to the regular flow over smooth models that had of necessity been the basis of his estimates.

We had no reliable information, based on sustained measurements, of the way in which velocity varied with height over a ridge. We did not know the height at which the maximum velocity would be found, nor the variation, if any, of this height with velocity. We thought that a sharp ridge would produce a high value of the maximum velocity at a low elevation; while a less sharp ridge would produce a lower value of the maximum velocity at a higher elevation. These assumptions are shown schematically by the dimensionless curves of Fig. 14.

In sum, we felt that the higher and sharper the ridge, the greater was likely to be the acceleration of the wind-flow over the summit; but we also felt that such a ridge would induce heavy and rapidly shifting turbulence on the lee side, which might even catch a turbine aback and wreck it. We lacked criteria for evaluating these and other factors and for making a rigorous selection of a wind-turbine site on a ridge.

As will be described, these qualitative assumptions have been confirmed; however, our inability in 1939 to make quantitative interpretations of the effect of topography upon wind-flow has also been confirmed by six years of ex-

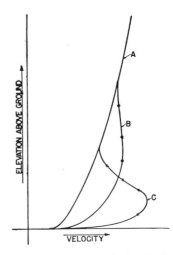

Fig. 14. Assumed velocity distribution above different types of terrain.

Curve A   Flat country.
Curve B   A rounded ridge, corresponding roughly to Curve A in Fig. 12.
Curve C   A sharp ridge.

perience. We selected as the test site a ridge which we thought possessed a speed-up or acceleration factor of at least 1.2. Our records indicate that this ridge apparently retarded the flow, giving a speed-up factor of only about 0.9. We have found no criteria by which to make an economically useful quantitative prediction of the effects of topography upon wind-flow.

3. *Prevailing Wind Direction at 2000 Feet in the Western Foothills of the Green Mountains.*

Upper-air data indicated a prevailing wind direction of W by N in New England at 2000 meters (Fig. 15). This was confirmed at Mount Washington at

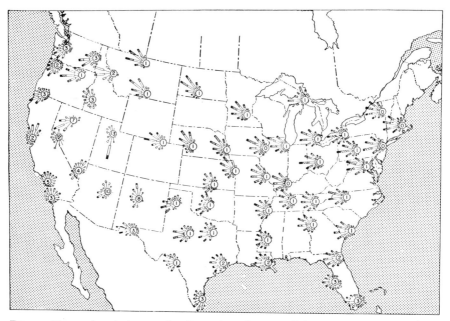

FIG. 15. Upper-air wind roses for December, January, February, 2000 meters above sea level, United States Department of Commerce Weather Bureau. Numbers in circles refer to the percentage of calm (0–3 meters per second). The length of the spokes indicates the steadiness of the wind.

6300 feet (Fig. 16A), and in the free-air over Burlington at a little over 4000 feet (Fig. 16B). It was therefore assumed that the prevailing direction at Grandpa's Knob at 2000 feet would be about W by S, being shifted about 20 degrees to the left from the direction observed at Burlington at 4000 feet, in conformity with the Ekman spiral (Fig. 17). The error in this assumption, amounting to some 28 degrees (Chapter IV), presumably contributed to the error in the predicted output at Grandpa's Knob.

As will be described in more detail in Chapter IV, these first three assumptions of 1939 were in error and, although we have since located a 4000-foot ridge in Vermont where the originally predicted output of 4400 kilowatt-hours per kilo-

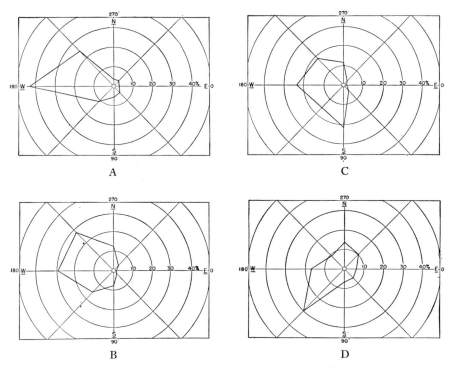

FIG. 16. Directional distribution of wind in New England. Five year averages.

A—Mt. Washington anemoscope,   6300 feet
B—Burlington pilot-balloon data,   4270 feet
C—Burlington pilot-balloon data,   2440 feet
D—Grandpa's Knob anemoscope,   2000 feet

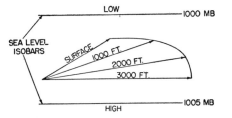

FIG. 17. Directional shift of wind in the Ekman spiral. From Petterssen's "Introduction to Meteorology," Fig. 67, page 108.

31

watt per year could probably be obtained, at the 2000-foot test site we averaged only 1200 kilowatt-hours per kilowatt per year, or 30 per cent of the output originally predicted for this elevation.

4. *Influence of the Structure of the Wind on Design.*

We assumed that, if the aerodynamics of a crest were such as to accelerate the wind-flow, the work done in compressing the streamlines would tend to damp out turbulence. Conversely, if the crest retarded the wind-flow, turbulence might be increased. We thought we were safe in designing the test unit, which was to be located on a site with an acceleration factor of not less than 1.20, according to the gustiness factors established over fairly flat country, as at Lakehurst, N. J., and later confirmed and amplified by Sherlock (as noted by Petterssen in Ref. 2-A). Our conclusions, based on this work, were that on well-exposed, fairly flat sites, the net velocity integrated over a disc area of 175 feet in diameter might change at the rate of 50 per cent each second, throughout an interval of 1.6 seconds, while the net direction might simultaneously change at the rate of 90 degrees in 1.0 seconds (Ref. 3-B). As will be explained, our experience with these two assumptions was inconclusive, but we have found no reason to question them.

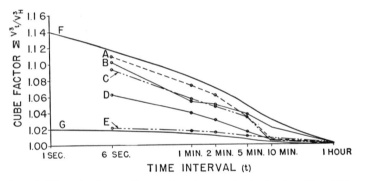

Fig. 18. The cube factor. From 5 hourly records obtained at Blue Hill Observatory, selected as representative by C. F. Brooks.

	Average Hourly Velocity	*Stability Conditions*
Curve A	Velocity 29.0 miles per hour	Very unstable
Curve B	Velocity 18.7 miles per hour	Unstable
Curve C	Velocity 39.1 miles per hour	Unstable
Curve D	Velocity 47.0 miles per hour	Unstable
Curve E	Velocity 17.8 miles per hour	Stable

5. *Influence of the Structure of the Wind on Estimates of Output.*

The power in the wind varies as the cube of the wind velocity (Chapter V). An anemometer records velocity. Therefore, to compute potential turbine outputs from anemometer records, it is necessary to deal with the cube of the values recorded by the anemometer. But anemometer records are frequently given in terms of total miles of wind per hour, and the wind does not blow steadily at this average value throughout the 60 minutes. It will, on the contrary, have been fluctuating, as shown by the gustiness patterns reported by Sherlock. The cube root of the mean of a series of cubed terms is greater than the mean of the series. For example, the cube root of the mean of the cubes of the series 20.0, 30.0, 40.0, is 32.1, while the mean of the series is 30.0. The ratio 32.1/30.0 we have called the cube factor. It is always greater than 1.0.

In Fig. 18 five examples of the cube factor are plotted against the logarithm of the time range of 6 seconds to 1 hour. They were derived from records taken at Blue Hill, selected as representative by C. F. Brooks. The range of mean hourly velocities covered is from 17 miles an hour to 47 miles an hour. By extrapolation of the arbitrary envelopes it is seen that the annual average cube factor in the range of 1 second to 1 hour lies within the range of 1.02 to 1.14, perhaps not far from 1.08, while the average for the range of 0.1 second to 1 hour is probably a little greater, perhaps 1.10.

We did not know what minimum unit of time was characteristic of the response of the wind-turbine, and accordingly we ignored the cube factor, treating it as a safety factor in estimating outputs from mean hourly velocities.

6. *Influence of Atmospheric Density on Estimates of Output.*

Output in kilowatts varies directly with the density (Chapter V), which in turn varies with the temperature and the elevation above sea level. The average sea-level density in New England in January is more than 10 per cent higher than in July. To obtain a value for the mean annual density at sea level, the average for each month was computed from the 5-year average temperature for each month, and from these monthly densities the annual average was computed by weighting the density for each month according to the computed output for that month (Ref. 4-B).

Density also varies with elevation (Fig. 19), the annual average value in New England at 10,000 feet being nearly 30 per cent less than at sea level (Ref. 5-B).

From these considerations, we estimated that the average annual weighted density at Grandpa's (2000 feet) would be 1195 grams per cubic meter. This was a good estimate. The actual weighting, based on our records, yields a 5-year average of 1201 grams per cubic meter, or half a per cent higher.

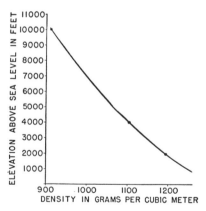

Fig. 19. Variation of atmospheric density with elevation. Annual values weighted for mean monthly temperatures in New England.

7. Influence of Estimates of Icing on Design and on Site Selection.

Brooks surveyed available data over a 35-year period and prepared charts show-ing how the maximum icing would vary with latitude and elevation above sea level (Figs. 20 and 21) (Ref. 6-B).

Brooks recognized that the deposit of an "ice storm" ranges in habit and density from light coatings of hoarfrost through rime and porous ice to clear, solid ice.

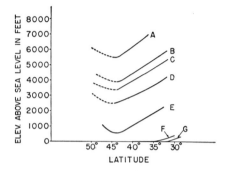 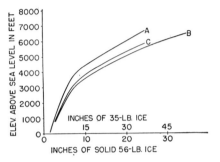

FIG. 20. The maximum thickness of 35-pound ice to be expected on a stationary structure in various latitudes in U. S. A.

Curve A	36.0 inches
Curve B	18.0 inches
Curve C	13.5 inches
Curve D	9.0 inches
Curve E	4.5 inches
Curve F	2.5 inches
Curve G	1.0 inches

FIG. 21. The maximum thickness of ice to be expected on a stationary structure in New England. Derived from Fig. 20 by Brooks.

Curve A	Latitude 58° 00′
Curve B	Latitude 43° 45′
Curve C	Latitude 47° 00′

Using the last as a standard, he showed that in Vermont the maximum thickness of solid 56-pound ice which might accumulate on a stationary structure, increased rapidly with elevation above sea level, from 5 inches at 2000 feet to 12 inches at 4000 feet.

No authority consulted would predict how ice would form on our stainless-steel blades, under conditions of maximum icing. We took the view that the test unit must not fail structurally; that if it did, the project would die.

This philosophy, in conjunction with our fears of heavy ice, became a strong incentive to create a design with much excess strength and to locate the test unit on the lowest site possessing wind adequate for the test program.

First Attempts at Answering the Questions and Testing the Assumptions

Question 1. Of the two practical questions to which we sought answers, the first, being the selection of the test site, was the most pressing. Certain require-ments had been set up. The site should have a clear sweep up-wind (thought to be

W by S), and a clean exit or tailrace to the east. It should have a bold and rounded profile. It should be located near the load center of the Central Vermont Public Service Corporation. It could not be located within the boundaries of the Green Mountain National Forest, and, most important of all, in order to avoid heavy icing, it should not be more than 2000 feet above sea level (Figs. 20 and 21).

It was necessary to select the test site in the spring of 1940. During the preceding winter, we had made map studies of Vermont, and, using certain criteria since found to have been unreliable, we had identified some 50 summits which we then thought were promising wind-turbine sites. Concurrently with the map studies we had made field trips, which failed to shed much light on the problem. Following a suggestion of Brooks, S. Morgan Smith Company authorized us to invite Griggs, the ecologist, to accompany us. Griggs could find little evidence of wind below 2500 feet (Ref. 7-B) but the meteorologists concluded that the wooded summits acting as airfoils would show compressed streamlines above tree-top height.

This view received some support when, during the early part of 1940, the results began to come out of the wind-tunnel at Akron (Ref. 8-B), showing that the models of Pond Mountain and Glastenbury, for example, would yield acceleration factors of 1.20, with the maximum compression of the streamlines occurring at higher elevations above the summits as the roughness increased.

Grandpa's Knob met the requirements of elevation, exposure, and profile, and was close to Rutland (12 miles), requiring only 2.2 miles of new road to connect it to a highway network. Accordingly, although the trees showed no deformation by wind and although there had not been time to put a model of Grandpa's into the wind-tunnel, it was selected as the test site in June, 1940, without benefit of anemometry.

An anemometer mast had been put up a few days before this decision was made and anemometry had begun immediately. The wind velocity records throughout the summer of 1940 gave computed monthly outputs which averaged only 10 per cent to 30 per cent of the predicted values. Unwilling to believe that these low values were characteristic of the western foothills of the Green Mountains, Wilcox studied the trend of the reported velocity records from Blue Hill, East Boston, and Mount Washington (Ref. 9-B), and discovered that, beginning a few months earlier, the velocity on Mount Washington had shown a substantial drop. From this we took heart and referred to the weak wind-flow at the test site as part of a temporary but regional "anomaly." It was not until the summer of 1945, in the course of the rigorous analysis that led to the abandonment of the project by S. Morgan Smith Company, that it was learned that the "anomaly" at Mount Washington had been caused by the application of an arbitrary correction to the anemometer records. The correction had been applied by one of the observers without notification to the users of the published data. It is quite likely that we have this observer to thank for the Smith-Putnam Wind-Turbine experiment. If it had been known at the end of 1940 that not only was there no anomaly, but also little wind at those elevations below which we did not fear ice, it is likely that the experiment would have been abandoned out of hand.

With the test site selected, we were free to develop programs to answer the remaining question in a fairly orderly manner.

Question 2. The Central Vermont Public Service Corporation requested that it be supplied at an early date with a list of the best potential wind-turbine sites in Vermont, each capable of development to 20,000 kilowatts of capacity, in order that the land might be placed under option.

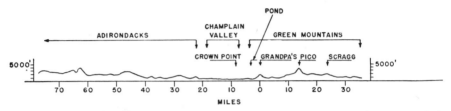

FIG. 22. East-west profile through the test site at Grandpa's Knob, 30 miles down-wind and 70 miles up-wind. Vertical scale exaggerated about 5 times.

The very best sites were, we thought, to be found among the high ridges of the Green Mountains. As early as 1935, in a report to the General Electric Company, I had picked the 3-mile Lincoln Ridge, trending N by E to S by W, as possibly the best site in Vermont (Ref. 10-B) on the grounds that it was unusually smooth and regular, with high relative elevation, a magnificent clear sweep across the Champlain Valley 20 miles to the Adirondacks (Fig. 22), a clean tailrace down-

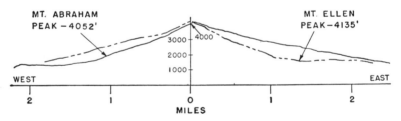

FIG. 23. West-east profiles through Mt. Abraham, the southern shoulder of Lincoln Ridge, and Mt. Ellen, the northern shoulder.

wind and a remarkably fair profile which looked promising aerodynamically (Fig. 23). However, this and most similar sites lay in the Green Mountain National Forest. It was several years before permission was obtained to consider sites in the Forest and before we had learned not to fear the heavier icing to be expected at 4000 feet. Accordingly, these sites were not surveyed in 1940–1941.

Lange thought that his experience in gliding would enable him to make a preliminary selection of ridges with good aerodynamic characteristics and capable of speeding-up the free-air velocity by a factor of at least 1.20. However, he lacked quantitative criteria for selection by this method alone, and particularly he lacked knowledge of the exact elevation over the summit at which the maximum acceleration of velocity would occur. Accordingly it was decided to check these tentative selections by testing in the wind-tunnel models of four selected mountains.

The Wind-Tunnel Tests

Topographic survey sheets of Mount Washington (6288 feet), Mount Glastenbury (3840 feet), East Mountain (2200 feet) and Pond Mountain (1407 feet) were blown up by pantograph and, from the "blow-ups," wooden models were made, layer upon layer, each conforming to a contour line. The outer corner of each layer was smoothed off and the model faired.

The models, to a scale of 1 foot to the mile, were 4 feet in diameter. They were mounted vertically on a ground-board in the vertical wind-tunnel of the Guggenheim Aeronautical Institute of Akron, Ohio, where the test program was carried out by the Director, Dr. Th. Troller, under the direction of von Kármán (Ref. 8-B). The models were rotatable through 180 degrees in steps of 45 degrees each. In this way, the direction of air-flow with respect to the model could be set from the North, Northeast, East, Southeast, South, Southwest, West, and Northwest.

A tunnel velocity of 85 miles an hour was chosen, in consultation with von Kármán. Some check readings were also made at 60 and 110 miles an hour.

The vertical distribution of velocity over the four mountain tops was measured at each of 80, 103, 113, and 127 points, respectively, in each of six wind directions, and at each of eight elevations above each point. Each measurement was made twice, and where the results did not coincide sufficiently closely, the measurement was made a third time.

In all, some 20,000 measurements were made twice by means of a hot wire anemometer consisting of a platinum wire 0.0006 inch in diameter, stretched and soldered to two needles. The wire was 0.25 inch long, this being the scale length of the radius of the turbine blade.

The minimum scale height at which measurements were taken was 0.075 inch or 30 feet. Above about 2.5 inches or 5000 feet, measurements showed normal velocity.

Measurements were made with the hot wire in the horizontal plane, as it was found that the apparent air-flow was not far from being normal to the length of the hot wire.

Artificial roughness was built up by cementing, with lacquer, quartz sand grains 3/32 inch in diameter, to strips of scotch tape, which were then fastened to the smooth surface of the model. It was assumed that quartz grains of this diameter would simulate full-scale forests of 40-foot trees. Coarse grains and cotton were used to simulate trees up to 100 feet in height. To simulate progressive de-timbering of a site, successive strips of scotch tape were peeled off.

37

In subsequent tests small tufts of thread, at a height of 0.25 inch above the model's surface, corresponding to about 125 feet above the surface in nature, were secured with sealing wax to pins which had been driven into the model. The tufts, photographed in each of five directions of wind, gave some indication of the effect of topography on local flow.

The wind-tunnel tests did not reproduce findings already made in full-scale. For example, the acceleration factor measured on Mount Washington full-scale

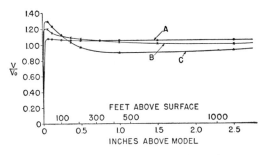

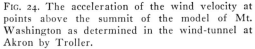

FIG. 24. The acceleration of the wind velocity at points above the summit of the model of Mt. Washington as determined in the wind-tunnel at Akron by Troller.

Curve A	Northwest Wind
Curve B	West Wind
Curve C	Southwest Wind

was about 1.50. But the wind-tunnel showed only 1.30 on Mount Washington (Fig. 24), while indicating 1.29 for Pond and 1.44 for Glastenbury.

Again, the balsams on Pond Mountain showed striking reverse flagging * just in the lee of the summit, indicating standing waves of turbulence in the prevailing west or southwest wind. We were not able to reproduce this effect in the wind-tunnel.

Thus we were not prepared to accept at face value the indication from the wind-tunnel that, over such bare summits as Pond and Mount Washington, the maximum velocity would be found at a height of 35 feet, with a rapid decrease in velocity above this height, in the case of certain wind directions. Such a distribution of velocity would, of course, be disadvantageous to the operation of a great wind-turbine, whose disc would extend from 40 feet to 225 feet above the ground.

We concluded that there were two factors which made it impossible to explore for wind-power sites by means of scale models in wind-tunnels.

The first is compressibility. In the wind-tunnel, the compressibility of the air is without importance, but in the flow of air across mountains it must be taken into account, especially in connection with the stability of atmospheric stratifica-

* Flagging is described on p. 54.

tion. It is true that the effects of compressibility can be estimated qualitatively by the principle of dynamic similarity. But our consultants came to doubt whether extrapolation to full-scale in nature could be made with quantitative certainty.

The second is that, with the exception of mountains like Mount Washington, which are high relative to the country round about, we are probably uncertain of the true free-air velocity in nature. This uncertainty makes it impossible to relate with confidence a speed-up factor measured in the wind-tunnel to the factor to be found over the actual summit in question.

Because of these uncertainties we next attempted to measure the height and value of the maximum velocity in full-scale.

We tried two methods. First, we added 30-foot extensions to the 80-foot anemometer masts on those mountains whose models were in the wind-tunnel. Unfortunately those mountains were well timbered and, although the masts extended to 110 feet, they gave us a net anemometer height above tree top of only 60 to 80 feet. Readings showed us that the height of the maximum velocity was above this level. This result had been anticipated, and we immediately tried the second method, floating balloons.

We undertook to float a sufficient number of balloons over the summit of Pond Mountain (whose model was in the tunnel) to establish the vertical distribution of velocity over the summit. This velocity distribution was to be compared with the various velocity distributions obtained in the wind-tunnel, at the various roughnesses of the model, in order to find that scale of roughness which corresponded to nature.

The method failed. It proved impossible in the time available to fly enough balloons exactly over the point on the crest to establish the vertical velocity distribution.

A secondary objective of the floating-balloon runs was to correlate the rupture of the stream flow with the geometry of the profile. We wished to know how steep a ridge could be without producing turbulence near the summit which would interfere with the flow through the turbine disc. For this purpose, we tried to float balloons over Pond Mountain, Biddie Ridge, and other sites. We failed in this objective also. Most of the runs indicated turbulence in the lee, but this frequently caused the balloon to dip down out of sight.

The Necessity for Special Wind Research Programs

It had now become clear that theory was inadequate and that other methods had failed either to evaluate design decisions already made or to guide us in selecting sites for ultimate development. Accordingly, the consultants recommended to Beauchamp Smith that special wind-research programs had become essential. Smith concurred, and the coordinated meteorological and ecological studies described in Chapter III were launched. The findings are summarized in Chapter IV.

POWER FROM THE WIND

SUMMARY

1. A preliminary world survey of windy regions is summarized in Tables I and II.

2. We lacked general wind velocity data for the mountains of New England.

3. In order to launch the Project, our consultants in 1939–1940 made working assumptions, concerning:

  (a) The free-air wind velocity at mountain-top height in Vermont.

  (b) The effect of the geometry of a mountain upon the retardation or speed-up of wind-flow over its summit.

  (c) Prevailing wind directions in the western foothills of the Green Mountains.

  (d) Influence of the structure of the wind on design.

  (e) Influence of the structure of the wind on estimates of output.

  (f) Influence of the atmospheric density on estimates of output.

  (g) Influence of estimates of icing on design and on site selection.

4. In 1940, it became necessary to select Grandpa's Knob as the test site in the light of these assumptions, and before they could be tested by field work.

5. Attempts were made to test these assumptions by wind-tunnel tests of models of mountains and by floating-balloon runs. The tests failed.

6. To learn what we had to know about the wind-flow in mountainous country, we decided to launch special wind research programs, coordinating meteorology with ecology, as described in Chapter III.

Chapter III

THE SPECIAL WIND–RESEARCH PROGRAMS, 1940–1945

Introduction

In Chapter II the state of our knowledge concerning the habit of wind in mountainous country was found to be meager and uncertain. The working assumptions we had had to make in 1939 were described, and it was explained why we were uneasy about most of the important ones, how the first attempts to test them, by wind-tunnel and by floating balloon, had failed, and why we felt it necessary to launch special wind-research programs, in which we would attempt to correlate meteorological and ecological measurements.

The meteorology was placed in charge of Petterssen, assisted by Lange, with Rossby, Brooks, and von Kármán in consultation. The ecological correlation was carried out by Griggs, with collaboration by myself, and both programs were at first placed under my direction, since Wilbur, as Chief Engineer of the Project, was fully occupied with design problems until after the erection of the test unit was well under way.

The Meteorological Program

It was decided to take velocity measurements on potential sites both down-wind of the main mass of the Green Mountains and also up-wind. In order to investigate extremes, measurements were also to be taken both in the Champlain Valley at Crown Point, on Lake Champlain, and on certain 4000-foot summits in the main range of the Green Mountains. These geographical relationships are shown, partly schematically, on the profile of Fig. 22. Finally, in his report of March, 1940, Lange (Ref. 11-B) suggested equipping a number of un-suitable sites with recording anemometers. This was in a further attempt to explore the relationship between the geometry of the profile of a site and its effect upon the wind-flow.

As regards all these cases, Lange suggested that the speed-up factor could be determined by comparing the wind velocity at the site, as measured by anemometer, with the gradient velocity, as computed from the isobars on the daily weather maps.

In response to these various suggestions, anemometers were installed as shown in Table IV, at sites shown in Figs. 22 and 25, selected for the various reasons itemized in Table V.

TABLE IV. THE SMITH-PUTNAM EXPERIMENTAL ANEMOMETER STATIONS

Site	Anemometer Elevations			Period of Operation		Days of Operation
	Above Ground	Above Trees	Above Sea Level	From	To	
Pico Peak	40	10	4007	Jan. 31, 1941	March 24, 1941	52
	80	50	4047	July 2, 1940	April 30, 1941	301
	110	80	4077	Jan. 31, 1941	March 24, 1941	52
Glastenbury	80	40	3840	March 2, 1941	Sept. 30, 1941	212
Scragg	53	45	2633	May 31, 1940	Oct. 10, 1940	133
	66	38	2646	May 31, 1940	Oct. 10, 1940	133
	78	70	2658	May 31, 1940	Dec. 1, 1940	185
	108	100	2688	May 31, 1940	Oct. 10, 1940	133
Herrick	80	80	2640	June 1, 1940	March 20, 1941	292
Chittenden	80	30	2430	April 25, 1940	Nov. 30, 1940	219
East Mountain	80	40	2200	May 17, 1940	Aug. 10, 1940	85
Seward	59	13	2139	July 11, 1940	Feb. 2, 1941	206
	70	24	2150	July 11, 1940	Feb. 2, 1941	206
	80	34	2160	July 11, 1940	Feb. 2, 1941	206
	110	64	2190	July 11, 1940	Dec. 30, 1940	172
Biddie Knob Proper	75	61	2085	July 14, 1940	Dec. 1, 1940	140
Grandpa's Knob	50	50	2040	June 9, 1940	June 20, 1941	376
	64	64	2154	June 9, 1940	May 20, 1941	345
	80	80	2070	June 9, 1940	June 20, 1941	376
	110	110	2100	June 9, 1940	Dec. 18, 1940	192
	40	40	2030	Apr. 11, 1941	Dec. 31, 1945	1724
	80	80	2070	Apr. 10, 1941	Dec. 1, 1941	235
	120	120	2110	Apr. 1, 1941	Dec. 31, 1945	1731
	150	150	2140	Apr. 3, 1941	Dec. 1, 1941	242
	185	185	2175	Apr. 2, 1941	Dec. 31, 1945	1733
Biddie Knob I	50	4	1975	July 13, 1940	Dec. 31, 1940	171
	64	18	1989	July 13, 1940	Dec. 31, 1940	171
	80	34	2005	May 3, 1940	Dec. 31, 1940	242
	110	64	2035	July 13, 1940	Dec. 31, 1940	171
Moose Horn	76	46	1921	March 30, 1940	June 19, 1940	81
Middle	76	32	1836	Apr. 18, 1940	June 15, 1940	58
Pond	47	3	1407	Apr. 26, 1940	May 18, 1940	22
	64	20	1424	Apr. 26, 1940	May 18, 1940	22
	80	36	1440	Apr. 9, 1940	May 25, 1940	46
	110	66	1510	Apr. 26, 1940	May 18, 1940	22
Crown Point	151	—	240	Jan. 1, 1941	March 1, 1941	60

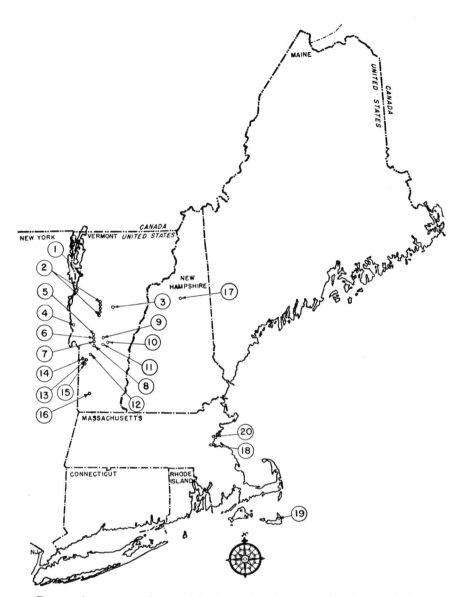

FIG. 25. Anemometer sites used in the project. See Table IV for description.

1. Burlington
2. Lincoln Ridge
3. Scragg Mt.
4. Crown Point
5. Biddie Knob
6. Seward
7. Biddie 1
8. Grandpa's Knob
9. Chittenden
10. Pico Peak

11. East Mountain
12. Herrick
13. Moosehorn
14. Pond
15. Middle Mt.
16. Glastenbury
17. Mt. Washington
18. Boston
19. Nantucket
20. East Boston Airport

43

TABLE V. THE REASONS FOR THE SELECTION OF THE VARIOUS ANEMOMETER STATIONS

Scragg

This potential wind-turbine site was interesting because it is in the wind shadow of the Green Mountains and its summit is bare. Private boundary lines restricted the location of the anemometer to a point other than that which would have been selected for a wind-turbine.

Herrick

The profile was thought to be inferior. The summit was bare and lay to the windward of the Green Mountains.

Chittenden

The site lay in the bottom of a horse-shoe, the open ends of which were directed downward and toward the west, which might, it was thought, accelerate the wind-flow by a funnel-effect.

East Mountain

Geographically a desirable test site, but so close to the Green Mountains (on their windward site) that the downwind exit was obstructed. The model was in the wind-tunnel.

Grandpa's Knob

Had been selected as the test site.

Seward
Biddie Knob Proper
Biddie Knob

Control stations on the same ridge as Grandpa's Knob.

Moose Horn
Middle
Pond

Selected to investigate the relationship between a higher peak (Moose Horn) down-wind of a lower ridge (Pond), and a gap (Middle), all up-wind of the higher Green Mountains.

Crown Point

To find an inland end point on Petterssen's curve of vertical gradient of velocity from the ground to the gradient wind level.

In order to measure the vertical distribution of velocity above wooded summits, three stations—Pond, Biddie, and Seward—were each equipped with three anemometers at various levels, while on Grandpa's there was erected the 185-foot Christmas Tree for the purpose of measuring the vertical distribution of velocity above a bare summit, the horizontal distribution of velocity, and the structure of the wind. The Christmas Tree, whose location is indicated in Fig. 26, is shown in Fig. 27.

By arrangement with the Department of Meteorology at Massachusetts Institute of Technology, special daily weather maps were prepared, from which

Willett, by standard methods, computed the gradient wind for the Vermont region for three periods each day.

Two purposes were to be served by using the gradient wind. First, it was hoped that it would serve as a source of an estimate of the velocity of the free-air at the elevation of each site, from which we could compute the respective speed-up

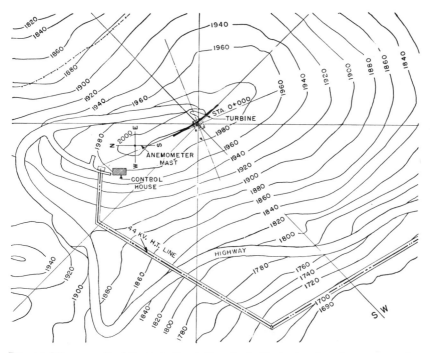

FIG. 26. The summit of the test site at Grandpa's Knob showing the access roadway, the 44-kv. high-tension line, the control house, the Christmas Tree anemometer mast, and the wind-turbine.

factors. In this way it would serve as a reference yardstick by which the sites could be compared with each other. Second, it was hoped that predictions of gradient wind velocity 30 hours in advance, with revisions 24, 18, 12, and 6 hours in advance, would make it possible to predict the output at each site, in a way which would be economically useful to a power dispatcher.

Instrumentation

The program of measuring wind velocity was begun with standard unheated Robinson $2\frac{5}{8}$-inch, 4-cup anemometers. Too many records were lost because of ice persisting in the cups and, in the winter of 1940–1941, we designed a heated anemometer known as the rotor type (Fig. 28), based partly on similar designs developed in Norway and at Mt. Washington. Lacking power lines at Pico,

FIG. 27. The 185-foot Christmas tree on Grandpa's Knob.

Grandpa's, and Glastenbury, we used gas heating in 1941. In 1943, we modified the design to accommodate an electric heating element, and later we incorporated other improvements. Five gas-heated anemometers were built, of which three were converted to electrical heating. After January, 1944, a gas heated and an electrically heated anemometer were simultaneously operated at the 120-foot level at Grandpa's.

All instruments were calibrated in a 1-meter wind-tunnel at frequent intervals, and usually showed no variation within the limits of the method (Ref. 12-B). In these tests, it was found that the table of correction for the Robinson instruments was essentially the standard table, which was accordingly used. In the case of the large rotors there was some suspicion of wall interference in this small tunnel, and accordingly each rotor instrument was calibrated by road test. The instrument was mounted well above a truck, which, at 4:00 A.M. in a dead calm, was driven in both directions over a 2-mile measured course. The truck was held at constant speed by speedometer, and the actual speed was determined by a stop watch. A calibra-

Fig. 28. The rotor-type heated anemometer.

tion curve was developed for each rotor, verifying the findings in the tunnel.

In conformity with our assumptions about the cube factor (Chapter II), it was found that the heavy rotors tended to overrun the lighter Robinson cup anemometers. Thus, the south rotor at 120 feet gave readings that in 44 months averaged 4 per cent higher than the lighter Robinson cup instrument on the west arm, at 120 feet. In Fig. 29 are plotted the ratios, north rotor/cup, and south rotor/cup, by month respectively, for the period November, 1941–June, 1945. Because of the interference of the Christmas Tree structure in the way of the prevailing wind, the north rotor recorded only about 95.2 per cent of the miles recorded by the better exposed south rotor. Monthly variations in this difference of 5 per cent between the two rotors are presumably due to varying directional distributions from month to month. Variations in the ratio south rotor/cup are presumably due to varying turbulence. The observed tendency toward a long-term trend in the two ratios, with a superimposed cyclical trend, has not been investigated.

An anemometer for use in conjunction with a wind-turbine should be dynamically similar to the turbine, as regards inertia and load. If unloaded and too light, it will presumably overrun by responding to the very short gusts of small diameter, which are without significance in turbine output; and if too heavy, it will also overrun.

Each indicated half-mile of wind was integrated and recorded by a stylus on the open scale waxed chart of a specially designed drum-type chronograph recorder (Fig. 30). The stylus was tripped by the closing of a contact in a direct current circuit powered with 6-volt dry cells. The drum was driven by clockwork. These recorders received daily servicing.

In December, 1944, at the request of the U. S. Weather Bureau, we added a USWB category-type integrating recorder at Grandpa's, on an experimental basis,

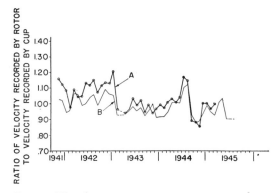

FIG. 29. The heavy rotor anemometers tend to record more miles of wind than the lighter Robinson Cup anemometers.

Curve A The ratio between the miles of wind recorded by the North rotor and by the cup anemometer at 120 feet on the Christmas Tree.

Curve B The ratio between the miles of wind recorded by the south rotor and by the cup anemometer at 120 feet on the Christmas Tree.

and in January, 1945, we added a demand-meter type (Fig. 31A) of integrating recorder which, each 15 minutes, prints the total mileage of wind passed, in arbitrary units (Fig. 31B).

The record of the category type is an accumulation of minutes of wind in each of the ten different categories of velocity shown. The limits of each category may be selected at will. We chose those listed in Table V.

The horizontal direction of the wind was recorded at Grandpa's for the period, April 9, 1941, to December 31, 1945, by means of a standard anemoscope, the wind vane of which was 30 feet above ground, on a special mast about 40 feet north of the anemometer mast. This instrument was serviced daily. A sock was rigged to the north cross-arm of the Christmas Tree at 120 feet, and vertically above the anemoscope, to determine the vertical distribution of wind direction over this part of the turbine disc. On Pico and Glastenbury, the resident observers tending the heated anemometers during winter months made certain visual observations of wind direction, according to a schedule.

The vertical direction of the wind was determined occasionally by measuring with a protractor the angle to the horizontal made by rime on the anemometer masts.

No regular aerological ascensions were made to determine the vertical distribution of temperature. The observer making the daily ascent on foot of Grandpa's

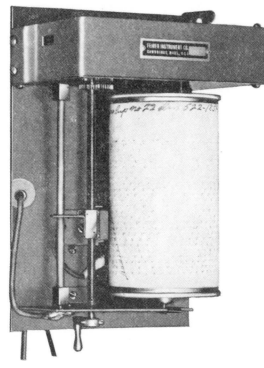

FIG. 30. The drum type recorder.

and Pico in 1940–1941 occasionally took thermometer readings at known points and times in the ascent. These records were useful only in indicating that strong temperature inversions (Chapter IV) did occur frequently.

The vertical distribution of velocity was determined by simultaneous hourly measurements on the Christmas Tree at Grandpa's, at 40 feet, 80 feet, 120 feet, 150 feet and 185 feet, from April, 1941, to December, 1941; and for periods varying from 1 month to 7 months during 1941, at three heights in the range 40 feet to 110 feet on Scragg, Seward, Biddie Knob, Pond; and at two heights, 35 feet and 97 feet, on Mt. Washington (Table IV) (Ref. 13-B).

At Grandpa's it was found that the anemometers at 40 feet, 80 feet, and

150 feet were in the zone of influence of the mast. They were then set out on struts 10 feet long, and therefore at a distance not more than 5 effective mast-diameters away from the mast. The interference continued so markedly at the 80-foot and 150-foot instruments that they were abandoned. The interference continued at the 40-foot instrument, probably causing the 5-year average to run low by 1 or 2 per cent in some wind directions.

FIG. 31A. The demand meter type of recorder.

At the other stations the masts were 5 inches in diameter and the two lower anemometers were on 3-foot struts, and thus at a distance of about 7 mast-diameters from the masts.

Instrumentation for the measurement of gust fronts—vertical and horizontal—had been developed by Sherlock who, however, informed Lange that his design was not suitable for our application. Accordingly, Lange designed a pressure-type anemometer which consisted of a pitot head faced into the wind by twin tail vanes. One of these anemometers was mounted, on a 10-foot standard, at the end of each of the four arms of the Christmas Tree at the 120-foot level; at the masthead at the 185-foot level; and at the 40-foot level; and the static head was mounted on a 10-foot strut 20 feet out from the mast on the cross-arm at the 120-foot level.

The choice of pressure instruments for the analysis of gusts was governed by the fact that we were not interested in the low velocities to which this type of instrument is insensitive.

Lange specified that the pressures were to be conducted down the mast to the 4-millimeter water manometers in the control house by ⅜-inch outside diameter tubes each 310 feet in length from the anemometer to the recorder.

The battery of manometers was mounted on the main instrument panel and

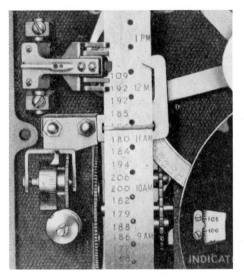

FIG. 31B. Close-up of the recording tape, demand meter type of recorder.

photographed by two cameras. One, to provide a continuous record, operated at 8 frames per minute. The other, to analyze critical gust data, operated at 8 frames per second. Approximately 750 hours of slow film and 150 minutes of fast film were exposed and are available. They have not been investigated.

The Ecological Program

Deformation by Wind.

Strong winds deform trees which, by the character and amount of deformation, then automatically integrate and record the force of the wind to which they have been subjected throughout their lives. Some species are more sensitive than others. Deciduous trees are poor indicators of wind velocity, showing little deformation other than brushing or stunting, probably because the amount of surface exposed to the wind varies greatly from summer to winter through the shedding of their leaves. Coniferous evergreens, which stand and take the weather the year 'round are, however, good indicators of the windiness of their habitats.

But deformation or injury caused by factors other than wind, such as temperature, insect injury, disease, lightning, or salt spray, is often mistaken by untrained observers for wind deformation. A particular caution should be given against mistaking for wind action the deformation of trees along seacoasts. Many books have striking pictures showing crippled trees exposed to the trade winds. B. W. Wells (Ref. 13-A) claims, and he supports his claim by chemical analyses, that crippling of trees along seacoasts is due largely to the salt in the spray deposited on them. The windward parts of the tree filter out the spray, leaving the leeward branches unwetted, and when the spray evaporates its salt is concentrated. Any young growth upon which it has been deposited is killed like grass on a salted tennis court.

Griggs has for years made a study of timber lines and the factors controlling their position, of which wind is one of the most important. As consultant to the S. Morgan Smith Company, he has developed his technique at various New England sites to a point where it is beginning to yield quantitative estimates of mean annual wind velocity within limits narrower than those of any method other than actual long-term anemometry (Ref. 14-B).

The correlation of the degrees of deformation occurring in a certain forest, with the wind and icing regime that has prevailed throughout the life of that stand of trees, is based on our present knowledge of the physiological responses that result in tree deformation. The physiology of the flagging of balsam will illustrate the process, in which five types of response are recognized.

1. The lateral branches just below the tips of the trunk may be bent to leeward by strong winds while they are still young, and held there until their tissues harden.

2. More often, however, a young branch bends into the wind rather than before it, especially if the wind is more or less intermittent, as in New England. If young balsams on a windswept hill are examined the day after a heavy blow, most of their twigs are found turned toward the wind. This tropism is probably due to differential transpiration. More water is abstracted by the drying winds from the windward side than from the leeward side of the twig. Growth is consequently greater on the lee side, and the twig curves into the wind. The mechanism by which this curvature is later reversed, finally setting the twig in a curve to leeward, has not been worked out.

3. For a similar reason the less exposed twigs on the lee side of the tree grow longer than those to windward, which lose more water through transpiration.

4. Such twigs as manage to grow to windward during the summer are killed by winter storms, leaving the trunk bare on the up-wind side.

5. In situations where winter storms create a heavy blast of snow and ice crystals, or sand, or even pebbles, the windward side of the trunk is abraded by the flying particles. The abrasion may vary from polishing the bark to deep erosion into the wood beneath. But such abrasion may be the record of merely a few

instances of very high velocity, and consequently its quantitative interpretation is obscure. Often the abrasion begins only some inches above the ground, and the base of the trunk is surrounded by a flat rosette of strong thrifty branches. Such "snow mats" indicate the permanent depth of the snow through the winter, and the wind-flow which permits snowdrifts to accumulate. Snow mats are widespread in flat lands, but in rough terrain the hollows catch most of the drifting snow and simple snow mats do not form.

However, snow is probably far more important as a protection from the severest winter storms than as an abrasive. Any thoughtful examination of trees on windswept situations makes clear the dependence of the trees on snowdrifts. Snowdrifts in turn are controlled by pockets of relatively slow air movement.

Balsam has been used as the example. Spruce responds similarly, but with certain important differences. Spruce requires good drainage and does not grow in places habitually covered with heavy snowdrifts, while balsam thrives under heavy snow, often recording in summer the location of winter snowdrifts by the profusion of new scrub balsam.

Correlation of Tree Deformation with Windiness.

It will have become clear that our understanding of the physiology and mechanics of the deformation of trees by wind was still too imperfect to permit a rigorous quantitative correlation between observed degrees of deformation and records of wind velocity. However, the area of our uncertainty was narrowed somewhat by recognizing that, while trees may be broken, they are not deformed by isolated storms, however violent.

In the New England hurricane of September, 1938, for instance, a velocity of about 200 miles per hour was reached on the summit of Mt. Washington. Great damage was done to tall trees in the forests of the lower and middle slopes; but the hurricane had no effect whatever on the deformed scrub trees with which we are dealing. This was attested by careful "before" and "after" studies in numerous locations. If the hurricane had affected timber-line trees, that effect would necessarily have been breakage or destruction. Brushing, flagging, throwing, clipping, carpets, and resurgence are adaptations gradually attained by living organisms under the long continued stress of severe conditions. The winds which produce deformation must be thought of as being not exceptional, but habitual.

Another evidence of this fact is that year after year the most severely clipped trees react in the same way to the stress of their environment. Over a period of ten years Griggs has followed the history of individual trees on Mt. Washington, by comparison of matched photographs. If their condition were due to special events, their vigor would vary from year to year as the winter was more or less severe. The almost complete uniformity of growth conditions from year to year is, however, one of the most remarkable features of these depressed trees.

Finally, it is manifest that a balsam or a spruce is very much more susceptible to injury during the few weeks of early summer, when its new twigs are soft and succulent, than after the new wood has ripened. Griggs has seen new growth completely blasted, but not deformed as described below, by spring winds of moderate intensity.

Types of Deformation Caused by Wind.

Brushing. A tree is said to be brushed when the branches are bent to leeward like the hair in a pelt which has been brushed one way. Brushing has been observed principally among deciduous trees, and is difficult to detect when they are in leaf. It is the most sensitive ecological indicator of air movement, giving clear indications of prevailing breezes too light to be of economic importance, and thus providing the lower end point of our ecological yardstick (Fig. 32).

Flagging. A tree is said to be flagged when its branches have been caused by the wind to stretch out to leeward, while the trunk is bare on the windward side, like a flagpole carrying a banner flapping in the breeze. From an economic point of view the most important indications of wind in New England are given by flagged balsam and spruce.

Throwing. A tree is said to be windthrown * when the main trunk, as well as the branches, is deformed so as to lean away from the prevailing wind, as though thrown to leeward. Throw is largely produced by the same mechanism that causes flagging; that is, the wind is strong enough to modify the growth of the more vigorous upright leaders, as well as the weaker laterals, which have a lower growth potential than the leaders.

Wind clipping. Trees are said to be wind clipped when the wind has been sufficiently severe to suppress the leaders and hold the tree tops to a common, abnormally low level. Every twig which rises above that level is promptly killed, so that the upper surface is as smooth as a well-kept hedge. Sometimes such "trees" occur with the needles of spruce, pine, and fir so intricately felted together that the individual tree is completely lost in the mass, which may be so dense that one can walk on the tree tops.

Tree carpets. Where the wind is so severe as to prohibit upright growth while still allowing trees to start on the ground level, every twig that reaches more than a few inches above the ground is promptly killed. The result is a living carpet of prostrate branches spreading over the ground. The many lateral buds which remain, after the destruction of all potential uprights, are too weak to form leaders, so all tendency to form erect trunks appears lost, and the whole carpet may be held to within 2 or 3 inches of the ground. It may, however, stretch out 100 feet or more to leeward of the sheltering rock where it originally started.

---

* The term "windthrow" is also sometimes applied to more violent damage as when a single storm so throws a tree as to break its roots on the windward side without actually uprooting it, thus bringing about another type of deformity, or more often the early death of the tree.

It is possible to differentiate between *loose carpets,* whose mats extend to about 18 inches above the ground, and *close carpets,* which, held to within a few inches of the ground, provide the upper end point on the yardstick of tree deformation by wind.

FIG. 32. A wind-brushed oak tree near the top of Blue Hill. The monument gives the location. Interesting because the wind regime on Blue Hill is well known.

Above this critical value of the mean velocity—about 27 miles an hour at specimen height—the tree cannot survive, and the rock remains bare, even though the site is well below timber line. Such a transition is found on the Horn of Mt. Washington, 1500 feet below timber line.

Winter killing and resurgence. Situations are often found where nearly all of a tree is kept clipped to a common level, but its central leader, by reason of its greater growth potential, manages to rise above that level during the summer. With the coming of winter the body of the tree is filled with a snowdrift, leaving only the new leader exposed. Certain winter killing is the result. In the following summer, however, the same conditions obtained as before and a new leader is sent up—usually in the lee of the first. This in turn is killed the next winter. So the process goes on year after year, indefinitely. As many as half a hundred such leaders have been counted on such resurgent trees.

The power of resurgent trees to continue putting out new leaders clearly depends on the vigor of the trees, as well as on the severity of the wind. From resurgence alone, therefore, it is not possible to get good quantitative estimates of wind velocity. Resurgence is, however, certain evidence that a site is subject to a heavy wind regime.

Deformation by Ice.

"Ice" may form on objects exposed to wind-driven moisture, over a fairly wide temperature range on either side of the freezing point. Brooks has pointed out that the deposits of an "ice storm" may range in habit and density from coatings of hoarfrost, through rime and light-weight aerated ice, to clear, solid ice weighing 56 pounds per cubic foot.

Deciduous trees suffer characteristic damage from heavy ice storms, and the condition of a deciduous forest along a potential wind-turbine site may serve as an indicator of the frequency and severity of heavy ice storms during the growing life of that stand of trees.

Where hardwood trees, such as beech, maple, or birch, heavily laden with ice, are subjected even to relatively moderate winds, great breakage results. The following year a thick brush of small branches is put out from the broken ends. If another ice storm with wind occurs the succeeding winter, these in turn are broken off. Naturally any branch rising above the general level of the forest is more subject to breakage from wind when iced than those branches within the boundary layer. A forest which is subject to ice storms, therefore, comes to be made up of characteristic even-topped, heavy-bodied, much-branched "candelabrum trees."

Despite the fact that spruces and balsams collect more rime than the deciduous trees, they are very much less subject to ice injury. In part, at least, this is probably because the relatively short branches of the spire-shaped evergreens, when ice-laden, exert less leverage than the long limbs of the diffusely branched hardwoods.

In 1939 a quantitative ecological yardstick did not exist. It was not possible to observe a stand of deformed trees and say: "On this site, at specimen height, the frequency distribution of wind velocities has been such and the maximum thickness of solid 56-pound ice has been such, during the life of these trees."

In an attempt to develop such yardsticks, Griggs carried out a program of field observation at each of the dozen or more of stations occupied from time to time by anemometers, and also at Mt. Washington, Blue Hill, and other sites of interest, such as Lincoln Ridge in Vermont.

The ecological evidence was analyzed and compared with the meteorological evidence collected by Petterssen, Lange, and Brooks. The conclusions were compared with the principles of dynamic meteorology, in collaboration with Rossby and von Kármán.

By 1945 there had been developed a crude but useful ecological yardstick for estimating windiness. It is described in the next Chapter.

SUMMARY

1. Special wind-research programs had been shown to be necessary.

2. A meteorological program was developed, in charge of Petterssen, with Lange, Rossby, Brooks and von Kármán in consultation. On 14 mountain tops, stations were established for measuring wind velocities; at 4 of these stations, the vertical gradient of velocity was measured; and at Grandpa's, the 185-foot Christmas Tree was erected, with 100-foot cross-arms at the 120-foot level, for measuring the structure of the wind.

3. New instruments were developed.

4. An ecological program was developed by Griggs, based on criteria described in this Chapter and correlated with the meteorological program.

Chapter IV

BEHAVIOR OF WIND IN THE MOUNTAINS OF NEW ENGLAND, 1940–1945

The Meteorological Evidence

Introduction.

The program of wind measurement described in Chapter III was carried out over the 5-year period, 1940–1945. Once the instruments were installed, the program was maintained by a very few individuals who were not called up for military service because, as a power plant, the project carried a double A priority rating.

However, it was not possible, because of the war, to maintain a staff adequate to evaluate all the measurements taken. The evaluation is still incomplete and the original data are available to any interested person. After V–J day, the data of most immediate interest—the direction and velocity measurements—were worked up by Lange, Wilcox, and myself.

The results, which we have not had time to analyze thoroughly, have been reviewed by von Kármán, Petterssen, Rossby, Willett, Brooks, and Lange and are here summarized, in preliminary engineering form.

The Frequency-Distribution of Wind Direction.

We have hourly observations of wind direction over the 5-year period, 1940–1945, from the anemoscopes on Mount Washington (Fig. 16A) and on Grandpa's (Fig. 16D). We have two series of pilot-balloon observations made four times a day at Burlington, throughout the 2-year period, 1941–1943, centered around the elevations 4270 feet and 2440 feet, respectively (Fig. 16B, C).*

The most frequent wind direction on Mount Washington (6300 feet) is a little north of west (about 285 degrees true). In the free-air over Burlington, at 4270 feet, the prevailing direction has shifted 5 degrees to the left and lies at about 280 degrees. A further shift to the left at Grandpa's (2000 feet), relative to that

---

* The total number of cases in this series was 3215, of which 3044 were used. To compute the velocity at 2440 feet, about 92 per cent of the 3044 were used; 77 per cent were used to compute the velocity at 4470 feet. The missing observations result from low cloud decks.

at 4270 feet, was to be expected, in conformity with the Ekman spiral (Chapter II), and amounting to about 20 degrees. This consideration alone indicated a value for the prevailing direction at Grandpa's of a little south of west (about 260 degrees), while the value actually found is 232 degrees (about southwest), a difference of 28 degrees.

It is possible that this differential shift of 28 degrees is caused by factors which are clearly operating in the case of the Burlington pilot-balloon data. In Fig. 16B it is seen that, while the prevailing direction in the free-air at Burlington at 4270 feet is in a rather broad band centered at about 280 degrees, the wind at 2440 feet (Fig. 16C) shows two bands of frequency accumulation, one centering at about 195 degrees, and one at about 130 degrees, with a minimum at about 235 degrees. In conformity with the Ekman spiral (Fig. 17), one would have expected to find the frequency-distribution at 2440 feet similar to that at 4270 feet, but shifted to the left and centered at about 260 degrees.

The explanation offered is that the mass of the Green Mountains, lying to the eastward of Burlington and averaging 4000 feet in height, tends to deflect the flow of the prevailing westerly winds below this elevation, so that the flow tends to be northerly and southerly around the ridge.

If this deflection is a fact, it should be reflected in the wind velocities recorded by pilot-balloon data.

Fig. 10 shows the velocity plotted against elevation, based on the mean of 2 observations daily during the 5-year period, June, 1940–July, 1945. At about 2300 feet the velocity is seen to decrease somewhat, dropping in summer about 4.5 per cent in about 500 feet before continuing to increase (Fig. 10). In 1939 it was assumed that this velocity decrease was apparent rather than real because, until the mountains had been passed, the balloon was being lifted in a rising wind, invalidating the assumption of uniform rate of ascent, and yielding, therefore, velocities lower than the true wind velocities. Accordingly, as described in Chapter II, this decrease in velocity was ignored in making our predictions of output at Grandpa's. Lange believes that part of this effect, and possibly most of it, is due to the drop in velocity accompanying the change in direction.

If this is the true explanation, the effect should be most pronounced in the relatively stable light winds of July mornings and least pronounced in the strong well-mixed winds of January afternoons. This is found to be the case in Fig. 10, where the curve for July mornings shows a pronounced loss in velocity, while the curve for January afternoons shows little effect.

There is no doubt that the effect of the Green Mountain range extends to the windward into the Burlington region, and causes the originally west wind to be deflected into southerly and northerly streams.

A wind rose of a similar bifurcated shape is indicated by the ecology on Mount Abraham (4100 feet), where the deformation indicates prevailing northwesterly winds, heavy enough to cause turbulence and reverse flow in the lee, both recorded by flagging of the spruces; and southwest winds, also recorded by flagged spruces, but unaccompanied by a standing turbulence severe enough to cause reverse flagging from the northeast.

The apparent deflection of the prevailing west wind at Grandpa's (28 degrees toward the south) is probably another example of the same type of influence.

The conclusion is that on the windward side of a mountain range there will be a dislocation in the distribution of wind directions as compared to that expected from the Ekman spiral; and a deficiency of wind velocity on the summits of the windward foothills, as compared with the velocity to be expected by interpolation in the standard vertical distribution of velocity from the ground to the gradient wind level.

Had this reasoning originally been applied to the height of the test site (1990 feet), it would have resulted in a decrease of 5 miles per hour in the prediction of the mean annual velocity, that is, from 24 miles per hour to 19 miles per hour. The value actually found at the test site was 16.7 miles per hour. If the value 19.0 miles an hour is the true speed of the free-air at the elevation of the test site, then Grandpa's Knob did not speed-up the wind, but rather slowed it down some 12 per cent, implying a speed-up factor of about 0.90.

The Frequency-Distribution of Wind Velocity.

The principal use of frequency-distribution curves of wind velocity is in the construction, for any turbine design, of a master curve to show the variation in

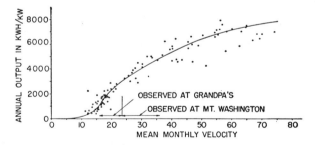

FIG. 33. The relationship between monthly output and the mean wind velocity for the month. For convenience in comparison with other output data referred to in the text, each monthly output has been converted to the equivalent annual output.

Each of the 120 points on the curve represents a monthly output computed from the anemometer records for that month, plotted against the mean velocity for the month.

This curve applies specifically to the dimensions and operating characteristics of the test unit.

annual output with variation in mean annual velocity, as shown in Figs. 33 and 34.

The many hundred velocity-frequency curves we have constructed have two origins, which stand in generic relation to each other. The first group was built up from actual hourly anemometer readings over periods varying from two to five years. Such frequency curves, for Blue Hill, Grandpa's, Nantucket, and Mount Washington, are shown in Fig. 35. The second group was built up from hypothetical curves, which can be constructed from an actual smoothed curve by using

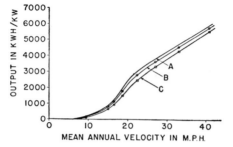

FIG. 34. Variation in annual output with mean annual velocity at three elevations above sea level.

    Curve A   Sea Level. Density equals 1254 grams per cubic meter.
    Curve B   4000 feet above sea level. Density equals 1105 grams per cubic meter.
    Curve C   10,000 feet above sea level. Density equals 911 grams per cubic meter.

These outputs were obtained by multiplying the eight velocity frequency-distribution curves of Table X by the power-velocity curve of Fig. 50.

This curve makes it possible to estimate the annual output from a turbine of this design at a site where the mean annual velocity is known, provided the wind regime has about the same velocity distribution as found in interior New England.

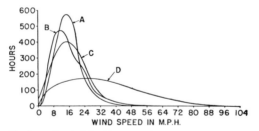

FIG. 35. Velocity distribution of wind in New England; 5-year averages.

    Curve A   Blue Hill, mean annual velocity 18 miles an hour.
    Curve B   Nantucket, mean annual velocity 16 miles an hour.
    Curve C   Grandpa's Knob, mean annual velocity 17 miles an hour.
    Curve D   Mt. Washington, mean annual velocity 34 miles an hour.*

Five-year averages. The area of each curve adds up to the 8760 hours of one year. The velocities are those recorded at the respective anemometer heights.

various assumed speed-up factors as multipliers. In Fig. 36 is shown the basic curve for Glastenbury as computed from pilot-balloon observations and pressure maps (speed-up factor = 1.00). To this curve there have been applied speed-up factors

* Uncorrected. See page 71; page 74; Fig. 43; and Table VII for corrected value. The frequency distribution curve of the corrected velocities is not available.

of 1.20 and 1.40 and the transformed and resmoothed curves are plotted on the same figure.

Experience gained in constructing velocity-frequency curves had taught us that all of the smooth curves were of a similar type, defined by statisticians as a

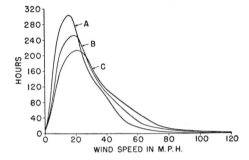

FIG. 36. Velocity distribution at Mt. Glastenbury (3840 feet).

Curve A Computed from pilot-balloon observations and pressure maps. It is a free-air velocity.
Curve B Results from applying a hypothetical speed-up factor of 1.20 to Curve A.
Curve C Results from applying a hypothetical speed-up factor of 1.40 to Curve A.

Pearson Type III function. We could, therefore, derive a frequency curve from a series of fixes such as occur when the Beaufort scale is used to report wind velocity at sea, as in Figs. 3 through 7. Curves were worked up in this manner for various oceanic locations, making use of the Weather Bureau's summary of reports from ships at sea.

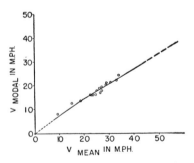

It is also possible to develop crude velocity-frequency curves when only the mean velocity is known, since the curves, at least those from the interior of New England, exhibit the following characteristics:

A. The intercept on the ordinate axis of hours is not at zero, but at some small number, usually at least 2 hours, and sometimes as high as 20 hours, representing no wind movement; that is, calms.

B. The two sides of the figure are concave upward.

C. The skewness increases with mean velocity.

FIG. 37. Relationship between the mean velocity and the most frequent velocity, in 16 velocity-frequency distribution curves.

D. The most frequent velocity is always lower than the mean velocity but varies with it (Fig. 37).

E. The number of hours during which the wind blows at the most frequent velocity decreases as the mean velocity increases (Fig. 38).

A curve sketched to meet these five requirements can be checked, since its area must total 8760 * hours and the mean velocity, computed graphically from it, must equal the value of the mean velocity taken as the fix.

Actual curves from a coastal station would doubtless show the effect of the diurnal sea breeze; and those from trade-wind stations would show a still different distribution. It is emphasized that the characteristics just described refer to curves of velocity-distribution found in the interior of New England.

Vertical Distribution of the Horizontal Component of Wind Velocity.

In Fig. 39 there has been plotted the average value of some 2500 hourly observations of velocities in excess of 15 miles per hour at Grandpa's, at 40 feet, 120 feet, and 185 feet, for the period, May 22, 1941, to December 1, 1941 (Ref. 15-B). All values are referred to 140 feet, the height of the hub of the test unit. From a study of the variation of this vertical distribution with wind direction, it was found that 85 per cent of all the hourly observations would fall within limiting curves, the band width being less than ± 3 per cent at 40 feet, and ± 1 per cent at 185 feet.

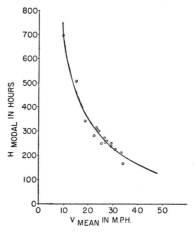

FIG. 38. The relationship between the number of hours during which the wind blows at the most frequent velocity, and the mean velocity, in 16 velocity-frequency distribution curves.

The number of hours during which the most frequent wind blows decreases as the mean velocity increases.

In Fig. 40 there are plotted the vertical distributions of velocity over three wooded mountains—Pond, Seward, and Biddie I—obtained from hourly records for the periods, April 27 to May 18, July 11 to November 30, and July 15 to November 30, 1940, respectively (Ref. 16-B). The ratios of velocities at the observed heights to the velocities at the standard height above the tree tops of 35 feet are plotted against the logarithm of the height above tree top. It will be seen that the conformity of the three distributions below the reference height is good. The dispersion of the points at 110 feet above the ground is thought to be due in part to the fact that these anemometers were mounted on 30-foot extensions of 2-inch pipe, which in high winds behaved like a whip lash, introducing unknown errors into the readings of these instruments. In each case the datum plane was the assumed average tree-top level. It is probable that these three distributions could be brought into greater coincidence by small changes in the estimates of the average height of the tree tops.

* The number of hours in a year is 24 × 365 = 8760.

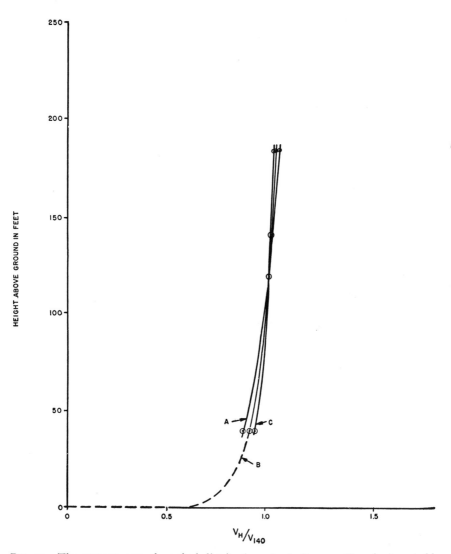

FIG. 39. The average annual vertical distribution of velocity over Grandpa's, a bald summit.

Curves A and C are arbitrary envelopes drawn around the mean value shown in Curve B. The envelopes include 85 per cent of the 2500 hourly observations, of which Curve B is the mean. The band width at 40 feet is ± 3 per cent, and at 185 feet is ± 1 per cent.

In Fig. 41 the mean vertical distribution of velocity over Grandpa's, a treeless summit, is compared with the mean distribution over the three wooded mountains. The vertical distribution over Mount Washington in the range of 35 feet to

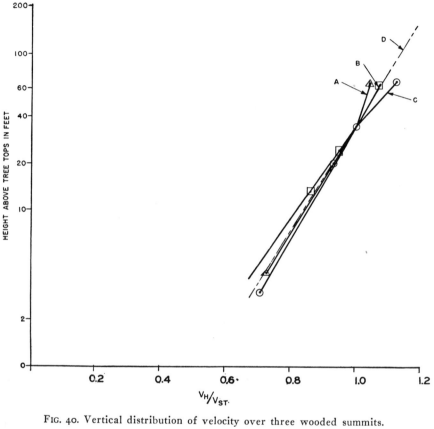

Fig. 40. Vertical distribution of velocity over three wooded summits.

Curve A	Biddie I
Curve B	Seward
Curve C	Pond
Curve D	The mean gradient

97 feet, obtained from hourly readings in the period, August 28 to September 13, 1940 (Table VI) (Ref. 17-B), is also plotted and found to be identical in slope with this sector of the Grandpa's curve.*

The important findings are that, on the average, the height at which the maximum velocity occurs over a summit such as Grandpa's is about 200 feet; and that the difference in average velocity between the lower edge of the turbine

* The summit of Mount Washington is cluttered with buildings, making the effective height of the two anemometers uncertain.

disc at 50 feet and the upper edge of the disc at 237 feet is not over 15 per cent. These results, which were not available until after the turbine was on the line, came as a relief to designers and backers alike.

It is interesting that the vertical distributions of velocity over such complex and varied airfoil profiles as those shown in Fig. 42 for the prevailing wind

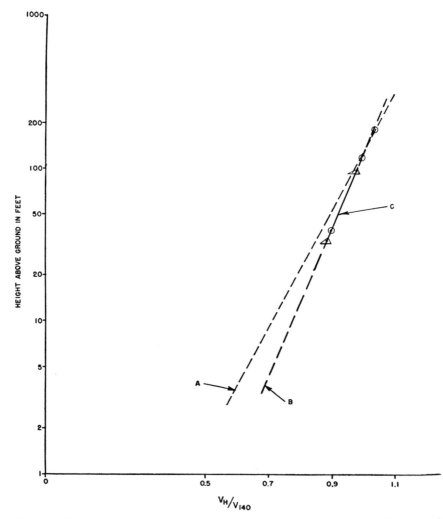

FIG. 41. The mean vertical distribution of velocity over Grandpa's, a treeless summit, compared with the mean distribution over the three wooded summits of Fig. 40.

Curve A   Mean distribution represented in Fig. 40 by Curve D.
Curve B   Mean distribution over Grandpa's.
Curve C   Observed distribution over Mt. Washington.

Compass Direction	Ratio	Per Cent
N	1.133	3.36
NE	1.623	3.60
E	0.957	2.90
SE	1.073	4.35
S	1.141	5.54
SW	1.185	10.93
W	1.058	44.02
NW	1.068	25.01
Mean	1.155	—
All, weighted	1.097	—
		100.00

The value of V_{mean} at 35 feet of 41.35 miles per hour has been supplied by Brooks, based on anemometry during the 5-year period July, 1941, through June, 1945.

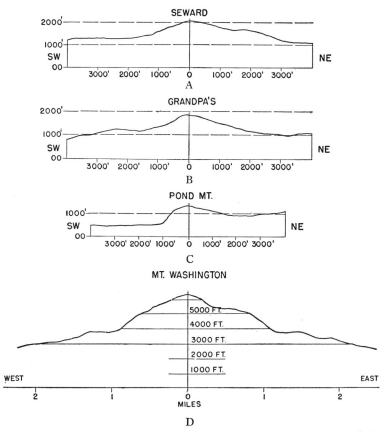

FIG. 42A, B, C, D. Profiles through four mountains drawn in the way of the prevailing wind at each mountain.

67

directions at these mountains, should not only be logarithmic, as predicted by Prandtl and as found by Rossby in the case of vertical distributions over flat lands, but should all have so nearly the same slope.

The Vertical Distribution of the Vertical Component of Velocity.

The angle of the wind-stream, over a ridge, measured from the horizontal, might be expected to vary in value from that of the slope of the ridge to zero degrees at some point near the height where the maximum velocity is found. At Mount Washington, Brooks reports a mean angle of 15 degrees in west winds, observed at a height of 35 feet. At Chittenden, I found 10 degrees at 50 feet (tree-top height being 30 feet) dropping to 3 degrees at 80 feet, in a west wind.

On Grandpa's the angle, as determined by rime on the Christmas Tree, has been observed to vary from 10 degrees at 20 feet to approximately 0 degrees at 185 feet. The average value over the disc area at Grandpa's is thus about 5 degrees, confirming the original estimates on which we had based the design angle of inclination of the turbine shaft.

Gust Fronts.

About 750 hours of film, at 8 frames per minute, and about 150 minutes at 8 frames per second, were taken of the 8 manometers recording the pressures in the pressure-type anemometers distributed over the Christmas Tree as shown in Fig. 27. These records have not been analyzed, and are available to any interested person.

If our assumption is correct that a compression of the streamlines tends to damp out gustiness, and if the speed-up factor at Grandpa's is actually less than unity (about 0.9), as we have concluded (p. 80, and Fig. 43), then it should follow that the gustiness factors we experienced during the test program exceeded what we would have found on Pico (4000 feet; speed-up factor $= 1.05$), or on Abraham (4000 feet; speed-up factor $= 1.7$). But, whether the maximum gustiness factor we did experience at Grandpa's exceeded the value we had assumed in designing the test unit, and described in Chapter II, is unknown.

Temperature Inversions.

That part of the atmosphere which concerns us is warmest at the bottom, where it absorbs heat from the earth's surface by radiation, conduction, turbulent exchange, and convection. The variation of temperature with height shows great fluctuations, but the normal rate is a decrease of 3 degrees Fahrenheit per 1000 feet of ascent.

Pressure decreases with elevation, and a gas, in expanding without addition or loss of heat, from one pressure to a lower one, suffers a drop in temperature. Thus a parcel of air moving upward suffers a drop in temperature as its pressure drops. Provided no heat is added or subtracted during the process of vertical movement, the temperature of the parcel of air falls almost exactly 1 degree centigrade per 100 meters (adiabatic lapse rate, or temperature gradient). This is 5.5 degrees Fahrenheit per 1000 feet, or nearly twice the normal rate.

When circumstances are such that the temperature increases with elevation, rather than decreases, we are dealing with an inversion, of which two kinds are of interest to us. These are the turbulence inversion and the radiational or ground inversion.

The turbulence inversion. The turbulence inversion occurs at the top of the layer of turbulence induced by the roughness of the surface. It is, therefore, more common and more pronounced over mountainous country. It marks the return, from the adiabatic lapse rate prevailing throughout the layer of turbulence, to the normal lapse rate above it. The wind velocity just above the inversion level tends to be higher than that just under it.

Since the inversion level is a layer of stability opposing passage of air through it, a turbulence inversion lying over the summit of a site, within some unknown critical height range, would act as one wall, though a yielding one, of a venturi tube, of which the other is a profile of the site. The streamlines that pass through the tube rather than around the ends of the ridge will be compressed, increasing the wind velocity over the ridge.

Radiational or ground inversions occur principally at night time, when a layer of cold air, cooled by conduction from the cold ground, underlies warmer upper air. Two principal and opposing effects of ground inversion enter into site selection. The first tends to increase velocities, by replacing the broken surface of the ground with the smoother, more frictionless surface of the top of the lake of cold, stagnant air. The second tends to decrease or obliterate wind movement. At a site drowned in such a lake there is little or no wind.

In addition, such layers of heavy cold air cascading down the sides of a mountain under gravity are common at night, often protecting the surface from the general wind.

Since no temperature lapse-rate measurements were made, we could not expect to correlate the ratio of wind velocity at Pico (4100 feet) with that of Grandpa's (2000 feet) in detail as a function of the temperature gradient.

But, the ratio of wind velocity at Pico to that at Grandpa's has been determined for day and night separately over a 7-month period, and has been found to be 4.8 per cent higher by night than by day, while that between Glastenbury (3700 feet) and Grandpa's, over an 18-month period, has been found to be 5.0 per cent higher by night than by day.

In the absence of temperature measurements, we do not know whether this indicated nocturnal stability is caused merely by a decrease of the lapse rate, or a lowering of the level of a turbulence inversion, or a ground inversion.

The net effect of inversion levels on wind-turbine sites in Vermont remains unknown.

Icing.

Although ice several inches thick was observed on the stationary structure several times, the maximum thickness observed on the rotating stainless-steel turbine blades was about ½ inch on the leading edge. As this skin of ice began to peel off, the unit would begin to run rough and was usually shut down for this reason.

Once or twice the unit was started up after the (stationary) blades had accumulated an unbroken sheet of ice about ¼ inch thick. Each of these times it was shut down before the ice had been flexed off, and for some reason other than roughness.

Grandpa's Knob was probably unusually ice-free during the test period (1940–1945), but the relative inability of ice to build up on the rotating blades gave us confidence that, if we encountered a heavy ice storm, we could either continue to operate or we could motor the blades at, say, 5 revolutions per minute, enough to cause flexing and to inhibit deep ice growth.

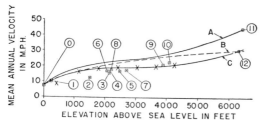

We concluded that ice was no bar to turbine operation at 4000 feet in the Green Mountains, although only further experience would tell whether blade de-icing equipment would be economically justified.

Fig. 43. The 5-year mean velocities observed and estimated at hub-height (140 feet), at the Smith-Putnam mountain-top stations, correlated with similar data for the same period at Mt. Washington, Blue Hill, and East Boston airport, and compared with the observed vertical distribution of the velocity of the free-air in the same height range (sea level to 6300 feet).

Curve B represents the original assumption of Curve B of Fig. 9.
Curve C represents free-air velocities.
Curve A represents accelerated velocities.*

Point o	East Boston	Point 6	Grandpa's
Point 1	Blue Hill	Point 7	Herrick
Point 2	Pond	Point 8	Seward
Point 3	Biddie Knob	Point 9	Glastenbury
Point 4	Biddie I	Point 10	Pico
Point 5	Chittenden	Point 11	Mt. Washington

Long-term Mean Velocities at Various Sites in New England.

In Fig. 43 are plotted the 5-year mean velocities as observed and estimated at the Smith-Putnam mountain-top stations (Ref. 18-B), correlated with similar data for the same period for Mount Washington, Blue Hill, and East Boston Airport, and compared with the observed vertical distribution of the velocity of the free-air in the same height range (sea level to 6300 feet). Computations are shown in Table VII.

The variation of free-air velocity with elevation above sea-level, in the range 434 feet to 4270 feet, was determined by Lange, who examined the four daily

* It is an arbitrary reference line, connecting the observed velocity at Mt. Washington with that at sea level at East Boston, and conforming to the free-air distribution of Curve C.

pilot-balloon runs at Burlington, 75 miles to the north of Grandpa's, from 1941 to 1945. He estimates a probable error in his curve as applied to the Burlington region of ± 2 per cent. But its application to the Grandpa region, 75 miles to the south, introduces an unknown error.

The value of the free-air velocity at the elevation of Mount Washington (6288 feet) was determined by Willett. He studied all North American upper-air data from 1910 through 1941, as summarized in upper-level pressure maps, which he analyzed in conjunction with pilot-balloon soundings from Burlington and Boston for the periods January, 1924–December, 1941, and November, 1926, through October, 1931, respectively.

He estimated the mean free-air velocity at the elevation of Mount Washington to be 30.0 miles per hour ± 10 per cent. (Point 12, Fig. 43.)

In Fig. 43, the lower solid-line curve through Lange's seven points was projected to pass through Willett's point for the elevation of Mount Washington, and through the sea-level value for the East Boston Airport.

The dotted-line curve is the assumption of 1939, discussed in Chapter II, and which resulted in over-estimating Grandpa's output.

The Long-term Mean Velocities at the Anemometer Stations.

The 5-year mean velocity on Mount Washington at anemometer height was supplied by Brooks. He determined a ratio between the velocity at anemometer height (35 feet) and the velocity at a temporary station at 97 feet on the Yankee Network Tower. This ratio is 1.10 compared with 1.08 for Grandpa's in the same height range (Fig. 39). The distribution of this ratio by direction is given in Table VI.

If this value is taken up to 140 feet by extrapolating along the Grandpa's curve of vertical distribution of velocity, we get a value for the mean annual velocity at 140 feet of 46.8 miles per hour. But the buildings on Mount Washington effect a net lowering of the heights of the anemometers, by an unknown amount; and there are other uncertainties. Accordingly, we arbitrarily take $V = 44.0 \pm 3.0$ as being the best judgment value for the 5-year mean velocity at 140 feet above the summit of Mount Washington. (Point 11, Fig. 43.)

The 5-year mean velocity at anemometer height on Blue Hill was also supplied by Brooks, and taken up to 140 feet along the curve of vertical distribution measured over Grandpa's.

The 5-year mean velocity at 120 feet on Grandpa's was recorded as described in Chapter II, and taken up to 140 feet along the measured vertical gradient of Fig. 39.

Of these three values of mean annual velocity at 140 feet, that for Grandpa's is the most reliable, while the values for Mount Washington and Blue Hill are based on the assumption that the vertical distribution of velocity over these two summits is identical with, or at least closely similar to, the vertical distribution

TABLE VII. CALCULATION OF V_{140} AT ANEMOMETER STATIONS

Station	1. Length of Period of Observation	2. Tree-Height, Feet	3. Anemometer Height Above Ground, Feet	4. Hub Height A.S.L., Feet	5. Horizontal Distance from Grandpa's, Miles	6. $\dfrac{V\,\text{anem.}}{V_{Gr_{80}}}$	7. Probable Error in Ratio of Col. 6
Crown Point	2 months	44	80	240	19.0	0.655	±13%
Pond	22 days	14	75	1501	13.0	0.693	±16%
Biddie Proper	5 months	50	80	2150	4.0	0.913	±0.6%
Chittenden	7 months	0	80	2490	10.5	0.894	±0.8%
Grandpa's	60 months	46	80	2130	0.0	1.000	±0.0%
Seward	6 months	0	80	2220	2.5	0.959	±0.5%
Herrick	9 months	46	80	2700	7.0	1.044	±0.4%
Biddie I	6 months	40	80	2065	1.5	0.909	±0.5%
Glastenbury	18 months	30	80	3900	46.0	1.152	±4%
Pico Peak	7 months		70	4107	13.5	1.235	±1%
Mt. Washington	60 months			6428	81.0		—

TABLE VII. CALCULATION OF THE MEAN ANNUAL VELOCITY AT HUB HEIGHT ($H = 140$ FEET) AT VARIOUS ANEMOMETER STATIONS

Station	8. $V_{anem.}$ Col. 6 × V_{Gr80}, mph.	9. Velocity at 140' With Trees — Gradient Factor, from Fig. 41	10. Velocity at 140' With Trees — Col. 8 × Col. 9, mph.	11. Velocity at 140' Trees "Removed" — Gradient Factor, from Fig. 41	12. Velocity at 140' Trees "Removed" — Col. 10 × Col. 11, mph.	13. $V_{Freeair}$, mph. from Fig. 43	14. Speed-up Factor Col. 12 × Col. 13, C_z
Crown Point	10.34	0.993	10.27	1.000	10.3 ± 15%		
Pond	10.94	1.125	12.31	1.036	12.8 ± 18%	18.0	0.71
Biddie Proper	14.41	1.089	15.69	1.010	15.9 ± 1.6%	18.9	0.84
Chittenden	14.12	1.143	16.14	1.043	16.8 ± 1.8%	19.0	0.89
Grandpa's	15.79	1.055	16.66	1.000	16.66 ± 0.0%	18.9 ⎫	0.88
Seward	15.14	1.130	17.11	1.038	17.8 ± 1.5%	19.0 ⎬ ± 5%	0.94
Herrick	16.48	1.055	17.39	1.000	17.4 ± 0.9%	19.0 ⎭	0.92
Biddie I	14.35	1.130	16.22	1.038	16.8 ± 1.5%	18.8	0.90
Glastenbury	18.19	1.116	20.30	1.033	21.0 ± 5.0%	20.2	1.04
Pico Peak	19.50	1.130	22.04	1.024	22.6 ± 1.5%	20.5 ⎱ ± 10%	1.10
Mt. Washington					44.0 ± 7%	30.0 ± 10%	1.47

over the summit of Grandpa's, at least up to 140 feet. It is true that all the vertical distributions we have measured support this assumption. Nevertheless, not enough vertical distributions above mountain tops have been measured to make the assumption secure and we must admit that the values of mean annual velocity at 140 feet at Mount Washington and Blue Hill may be in error either way by as much as 15 per cent.

Estimates of the 5-year mean values of velocity at 140 feet over Pond, Biddie Knob, Biddie Proper, Seward, Chittenden, Herrick, Glastenbury, Pico, and Crown Point involve uncertainties of another kind.

When we began our investigations of wind velocity in Vermont in 1939, we assumed that the only way in which to determine the long-term mean velocity at any mountain site was to maintain there a heated and tended anemometer for a period of a year or more. But we have learned that a year's record of wind velocity, standing by itself, is unlikely to be closely representative of the long-term mean velocity at that station. A convenient illustration of this is contained in the records of computed output for the test unit at Grandpa's. In analyzing five years of wind velocity records from Grandpa's, we find that the output computed from the velocity records for a single year may depart from the 5-year average of computed output by ± 26 per cent. In a 20-year period the departure of the computed output of a single year, from the 20-year average, would be greater—perhaps ± 30 per cent, as indicated in Fig. 44.

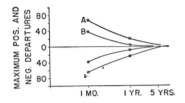

FIG. 44. Reliability of wind-power. Maximum positive and negative departures in output at two stations expressed as percentage deviations in output for the given time interval from the average output for that time interval experienced in five years of records.

    Curve A   Grandpa's
    Curve B   Mt. Washington

At Grandpa's, for example, the output in any single month may depart from the normal 5-year expectancy for that month by about 75 per cent.

What is needed is an estimate of the long-term mean annual velocity at the potential site. It is not practical to wait twenty years while this record accumulates.

Fortunately there appears to be a useful short cut. If the site in question is in the neighborhood of a long-established meteorological station, then the long-term mean velocity at the survey station may be determined as follows:

Determine the ratio between the velocity recorded at the survey station and that recorded at the control station in the same period of time. Multiply the long-term mean velocity at the control station by this ratio to obtain the long-term mean velocity at the survey station.

For example, let us assume that an anemometer station has been maintained for twenty years on a certain 200-foot mast. The long-term mean velocity at 200 feet is known to be 25.0 miles per hour. It is desired to find the long-term mean velocity at the 100-foot level on the mast. An anemometer is mounted there, on suitable struts to avoid mast shadow. It is operated for 24 hours, and the average of the 24 hourly values of the ratio between the velocity at 100 feet and that at 200 feet is found to be 0.8, with a maximum of 0.9 and a minimum of 0.6. At the end of 48 hours, the 48 observed ratios are collected into eight 6-hour averages, with a mean for the 8 values of 0.82, a maximum of 0.86, and a minimum of 0.74.

At the end of seven days, the average of the seven daily ratios is found to be 0.80, with a maximum of 0.81 and a minimum of 0.79, or the mean velocity at 100 feet is 80 ± 1.2 per cent of that at 200 feet = 0.80 × 25.0 = 20.0 ± 0.25 miles per hour; and so on, to any desired accuracy.

This method was used in estimating the 5-year average velocities at our various anemometer stations, whose records we ratioed in to the record at our control station at Grandpa's, where we had a continuous record for the five years, at the 120-foot level. The velocity found at the 120-foot level was converted to that at hub height, 140 feet, by interpolation on the curve of vertical distribution of velocity described on pp. 63–66 and in Fig. 39.

The length of the record required at the survey station in order to establish the ratio of the velocity there to that at the control station (Grandpa's, 140 feet), with a stated accuracy, varies with the slant distance between the stations (Fig. 45). Thus, to obtain the ratio with an error of about 10 per cent, it was necessary to operate Glastenbury, 38 miles away and 2000 feet higher than Grandpa's, for 90 days; Pico, 12 miles away and 2000 feet higher, for 30 days; Herrick, 10 miles away and 400 feet higher, for 10 days; and Grandpa's 185-foot station, 65 feet higher on the same mast, for about 14 hours.

In Table VII there are summarized the estimates of the 5-year mean values of velocity at 140 feet at our various anemometer stations, with some indications of the probable error in each value.

These estimates were made as follows:

First, there was computed, for each month of observation, the ratio between the wind movement at the observation station and the wind movement at the control station, viz., the 140-foot level at Grandpa's. The monthly values were averaged to give an average value for the ratio over the period of observations, which varied from 6 months to 18 months.

Next, the average value of the ratio, so computed for each observation station, was multiplied by the 5-year mean value of the velocity at Grandpa's at 140 feet.

All reliable observational material was used. When there was doubt concerning the condition of the instrument, particularly under icing, the record was thrown out. No attempt was made to select velocity readings by direction or by velocity group.

Sensitivity of this ratio to changes in wind direction. Since anemoscopes were not operated at the nine anemometer stations in question, it is not possible to make a direct analysis of variations in the ratio between the mean velocity at 140 feet at Station X and that at Grandpa's, with variations in the direction of the wind. There are a few clues, however.

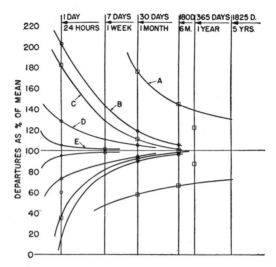

FIG. 45. A chart to indicate the minimum length of period of observation necessary to establish the ratio between the mean velocity at Station X and the mean velocity at the control station, with a given percentage of uncertainty.

Curve	Station
A	Departures in the ratio of the wind velocity at Mt. Washington to that at Grandpa's from the mean ratio.
B	Departures in the ratio of the wind velocity at Glastenbury to that at Grandpa's from the mean ratio.
C	Departures in the ratio of the wind velocity at Pico to that at Grandpa's from the mean ratio.
D	Departures in the ratio of the wind velocity at Herrick to that at Grandpa's from the mean ratio.
E	Departures in the ratio of the wind velocity at 185 feet on the anemometer mast at Grandpa's to that at 120 feet on the same mast from the mean ratio.

Explanation: For example, over the period of observation, which was about a year, the mean velocity in any 24 hours at the 185-foot station was never more than 6 per cent and never less than 5 per cent of the ratio between the mean velocity for the whole period at 185 feet and the similar mean velocity at 120 feet. But the minimum time interval required to establish a ratio with the same certainty between the wind velocity at Grandpa's and that on Herrick (4 miles away from Grandpa's and 500 feet higher), is 30 days.

At Grandpa's we do have a record of the variation in the ratio between mean velocity at 185 feet and that at 120 feet, with wind direction, for about 6 months (Table VIII). The mean value of this ratio, summing up and weighting all directions, is 1.042. The maximum deviation from this mean occurs in winds from the northeast and amounts to about 2 per cent. The records on the summit of

76

Mount Washington show much higher variations in the ratio between the velocity at 97 feet and that at 35 feet, as listed in Table VI. The ratio in an east wind is found to be about 59 per cent of the ratio in a northeast wind. This high variation, of course, is due to the many buildings which clutter up the summit of Mount Washington.

A third indication of a different sort was somewhat dubiously afforded by the wind-tunnel tests of model mountains. In these tests, whose uncertainties are

Table VIII

Variation in the ratio $\dfrac{\text{Mean velocity at 185 feet}}{\text{Mean velocity at 120 feet}}$, with variation in the wind direction, at Grandpa's, for about six months.

Direction	Ratio V_{185}/V_{120}
North	1.061
North East	1.064
East	1.048
South East	1.032
South	1.046
South West	1.040
West	1.029
North West	1.051
ALL	1.042

described in Chapter II, the velocity measured at a certain height above a certain point over the summit of the model was found to vary by as much as 40 per cent with changes in wind direction.

Finally, a clue was afforded by comparing the ratios of the mean velocity at 140 feet at the nine observation stations with that at Grandpa's for intervals of time shorter than one month. In the case of Pico, the ratio was determined for 6-hour periods and it was found that the dispersion was high (Fig. 45).

The explanation offered is that part of this dispersion is probably due to factors discussed on pp. 77–78, but that part of it is due to the directional distribution of wind velocity during the periods of observation.

Sensitivity of this ratio to changes in wind velocity. It has been found that, in the case of stations separated by a substantial slant distance, the ratio is not independent of velocity. Fig. 46 shows that the ratio between the mean velocity at 140 feet at Station X and that at Grandpa's is some function of the velocity at Station X. This means that ratios determined during periods when the wind velocity was below the long-term average will yield a ratio that is too low, and, similarly, if the ratio was determined during periods when the wind velocity exceeded the long-term average, the ratio will be too high. However, when the observation station lies at about the elevation of the control station and is exposed to a nearly identical wind regime, the uncertainty arising from this consideration becomes unimportant.

It will be noted that in Fig. 46 the ratio of the mean velocity at 140 feet over Seward to that at Grandpa's is nearly independent of the value of the velocity at Seward, whereas in the case of the comparison between Glastenbury and Grandpa's

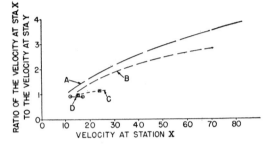

FIG. 46. The variation in the ratio between the mean velocity at two stations with the velocity at one of the stations. It is seen that the ratio increases as the velocity increases where the two stations are separated in elevation, but that where the two stations are not separated in elevation, the ratio is nearly independent of the velocity.

Curve A   Ratio between the mean velocity at Mt. Washington and the mean velocity at Grandpa's.

Curve B   Ratio between the mean velocity at Mt. Washington and the mean velocity at Glastenbury.

Curve C   Ratio between the mean velocity at Glastenbury and the mean velocity at Grandpa's.

Curve D   Ratio between the mean velocity at Seward and the mean velocity at Grandpa's.

there is a small but definite increase in the value of the ratio, with increasing velocity at Glastenbury.

A similar argument should apply to wind-velocity data coinciding with abnormal temperature inversions, which might produce extreme variations in the ratio between two summits separated somewhat in elevation.

If, therefore, the ratio were determined by using only those velocity records which coincided with:

A. The prevailing wind direction,

B. A wind velocity not far removed from the assumed long-term mean annual velocity, and

C. An absence of inversion levels,

then a convergence would be obtained substantially tighter than that of Fig. 45.

On the other hand, it must be admitted that the periods of observation which form the statistical basis for Fig. 45 were short, and that, had the experience been longer, greater aberrations would have been found, and the convergence would have been less tight.

Until more experience is gained, it is suggested that those two factors may balance each other. The convergencies of Fig. 45 may be taken as qualitatively representative of the variation in uncertainty of the ratio with the length of the period of observation, for various slant distances.

The uncertainties in the ratios so computed are tabulated in Column 7 of Table VII. Thus, the estimated long-term mean velocity at anemometer height is thought to contain a probable error at Pond Mountain of ± 16 per cent, at Crown Point of ± 13 per cent, at Glastenbury of ± 4 per cent, and at Biddie Proper, Chittenden, Seward, Herrick, Biddie I, and Pico of ± 1 per cent or less.

These ratios were first established between the respective anemometer heights. Values of the long-term mean velocity were then computed and these values were taken up to 140 feet along the curves of vertical distribution of velocity established in Figs. 39 and 40. In the case of bald summits (Herrick), the vertical distribution used was the average distribution measured at Grandpa's (Fig. 39). In the case of timbered summits, the distribution used was the mean of the distributions found at Pond, Seward and Biddie I (Fig. 40), followed by the "removal" of the trees by increasing the anemometer height by the amount of the height of the trees, along the Grandpa gradient. Thus all the velocities apply to bare slopes.

Table VII summarizes these computations, and in Column 12 of Table VII are indicated the over-all uncertainties in the velocities of Fig. 43.

From our limited experience in New England, we may conclude that observation stations which are separated from the control station by as much as 50 miles in horizontal distance and 2000 feet in vertical elevation, can be ratioed in to the control station, and the long-term mean velocity established, with an error of about 10 per cent, in a period of 90 days or less. If the field work is scheduled for the mean periods of April–May or September–October, and if care is used to exclude velocity data coinciding with abnormal temperature inversions or directions or velocities, it seems likely that the error would be about 10 per cent at the end of a 60-day period.

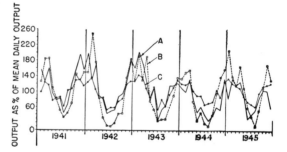

FIG. 47. Variation of the computed monthly output at three stations for 120 consecutive months, expressed as percentage deviations from the mean output for the five years.

In order to make the results comparable among months of differing lengths, the output for each month has been divided by the number of days in that month.

Curve A	Grandpa's
Curve B	Mt. Washington
Curve C	Blue Hill

Where stations are more closely related, whether in horizontal distance or in elevation, the time interval for a given degree of uncertainty is correspondingly reduced. Where stations are as widely separated as Mt. Washington and Grandpa's (say, 80 miles and 4000 feet), there is no useful relationship. This is shown in Fig. 45, and also in Fig. 47, on which are plotted the sixty computed monthly outputs at Mt. Washington, Blue Hill, and Grandpa's. It will be seen that there is no useful statistical relationship.

Mountains as Airfoils.

The upper curve of Fig. 43 is a reference curve, and joins the velocity actually observed on Mt. Washington (44 miles an hour) with that observed close to the ground at sea level (8 miles an hour). Thus the velocity at any point on the upper curve is related to the velocity at the underlying point on the lower curve by the speed-up factor C_x. At sea level this is unity, but on Mt. Washington it is $44/30 = 1.47$, which is the amount by which Mt. Washington, acting as an air-foil, has speeded up the free-air flow.

We had assumed in Chapter II that, in the case of aerodynamically identical mountain ridges, the speed-up factor C_x would tend to increase with the velocity. But our measurements show (Fig. 48) that while the speed-up factor at eight different sites does actually vary quite closely with the free-air velocity, it varies hardly at all with the geometrical differences among the eight profiles. This interesting and unexpected discovery has not been investigated.

At Grandpa's, whose location is shown in plan in Fig. 25 and in elevation in Fig. 22, we find that the speed-up factor C_x has a value of about 0.88 (Table VII).

At nearby Seward, whose elevation is almost the same as Grandpa's, the value of C_x is 0.94, or about 8 per cent higher. The respective WSW–ENE profiles are shown in Fig. 42. Although the Seward profile does appear to be somewhat fairer, it must be admitted that we lack criteria, based on map study alone, for predicting that the velocity over the summit of Seward would be 8 per cent greater than that over Grandpa's.

A spectacular exception to the relationship shown in Fig 48 is the Horn of

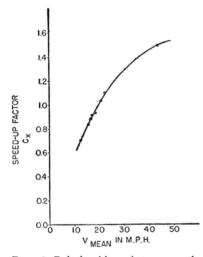

FIG. 48. Relationship between the speed-up factor, C_x, of a ridge and the free-air velocity over the ridge.
The eight values of the speed-up factor are taken from Column 14 of Table VII, and are derived from relationships shown in Fig. 43, and described on pages 70, 71, 74.
The amount by which a mountain ridge, acting as an airfoil, speeds up the free-air velocity, increases as the mean velocity increases. The low dispersion is remarkable in view of the wide differences in geometry among these mountain airfoils as shown in the profiles of Fig. 42.

Mt. Washington. The Horn, which is the northern end of Chandler Ridge, lies at 4100 feet, on the east-west profile shown in Fig. 49. Brief wind-velocity measurements at the Horn have confirmed the impression of the summit observers, that the mean velocity on the Horn is substantially greater than that on the summit, 2200 feet higher. Specifically, the mean velocity at 140 feet over the Horn was found to be 47 miles an hour, giving an apparent value for the speed-up factor there of about 2.10.

It seems clear that we know very little about the effect of the geometry of a mountain-mass upon the wind-flow.

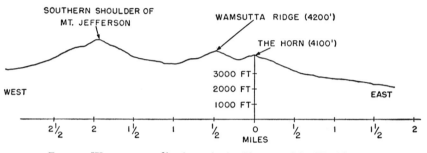

FIG. 49. West-east profile through the Horn on Mt. Washington.

SUMMARY OF THE METEOROLOGICAL EVIDENCE

1. *The Frequency-Distribution of Wind Direction.*

On the windward side of a mountain range, there will be a dislocation in the distribution of wind directions, as compared with the Ekman spiral; and a deficiency of wind velocity over the summits of the windward foothills, as compared with the velocity to be expected by interpolation in the standard vertical distribution of velocity from the ground to the gradient wind level.

2. *The Frequency-Distribution of Wind Velocity.*

The use for a velocity-frequency distribution curve is described.
Methods of constructing the curve are given.
Examples are shown.

3. *The Vertical Distribution of the Horizontal Component of Wind Velocity.*

a. At Grandpa's the average height at which the maximum velocity occurs is not over 250 feet. Turbine towers more than 200 feet high are uneconomical, at least on hilltops.

b. The curves of vertical distribution of velocity, over all the summits measured, are logarithmic, and with nearly identical slopes.

4. *The Vertical Distribution of the Vertical Component of Wind Velocity.*

The mean flow through the turbine disc at Grandpa's, extending from 50 feet above the ground to 237 feet, is *up* about 5 degrees from the horizontal.

5. *Icing.*

Icing was unusually light throughout the test period. The stainless-steel blades appeared to flex off light ice easily. It is not known whether de-icers would be economically justified at 4000 feet in Vermont.

6. A method has been developed for determining the long-term mean velocity at an observation station by ratioing the results of a 60-day period of observation to the simultaneous value observed at a control station, and multiplying the long-term mean velocity at the control station by this ratio. This method was developed in the course of estimating the long-term mean velocities at eight mountain stations in New England.

7. On the eight "average" sites studied, the speed-up factor depends upon the free-air velocity, very little upon the profile of the site.

8. A spectacular exception to this (the Horn at 4100 feet on Mt. Washington) makes it clear that we know very little about the effect of the geometry of a mountain-mass upon the wind-flow.

9. Seward (2080 feet), the best of the low stations, would have yielded 27 per cent greater output than Grandpa's, the test site (1990 feet), for reasons suspected, but not yet known.

The Ecological Evidence

The wind-power engineer is interested in the mean annual velocity which he can count upon for the generation of electricity. It was made clear in Chapter III that considerable components of the mean velocity do not contribute to the deformation of trees—the occasional very severe storms which, although of short duration, pile up considerable increments to the annual mean, and the light breezes which, because of their prevalence through many days, also contribute a large fraction to the whole.

On the other hand, it seems equally certain that for each species of tree in each habitat, there is some critical value of the mean wind velocity below which deformation will not occur, no matter how constant and sustained the wind may be.

This limit is obviously difficult to establish, and today we have only a general idea of where it lies in the cases of one or two species.

But wind-turbines likewise can make little use of light wind, and hurricanes are sources of danger rather than of power to the installations. Hence, in dealing both with trees and with turbines, it would seem, at first, that one should really be concerned with something other than mean annual velocity. But, as will be shown presently, this is not the case in turbine design. It has been found that, in interior New England at least, if we know the mean annual velocity, then, for any turbine design, we can know the output (Fig. 34). This is because the velocity-frequency distribution curves, measured in New England, are all of the same statistical type; we are permitted, to make the tentative assumption therefore, that, in New England at least, tree deformation is likewise a function of mean annual velocity. The problem, so far as the trees are concerned, is to ascertain at what value of the mean annual velocity the wind begins to flag balsams and spruces and how frequently a wind of given velocity must blow in order to be reflected in the shape of the trees.

The first step in correlating degrees of tree deformation with measurements of wind velocity was to relate the observation stations to each other in an ascending order of measured mean annual velocity, and to compare this list with a list of the same sites arranged in an ascending order of observed tree deformation. Would the lists match? At the first attempt, Griggs readily differentiated between the three groups of sites with high, moderate, and low mean annual velocities, respectively. Later work brought more refined correlation within the groups and, in fact, pointed to the necessity for a careful redetermination of the long-term mean velocity at each site.

The second step was to determine the quantitative relationship between the measured or estimated value of the long-term mean velocity at specimen height, and the degree of deformation.

In Table IX are set forth, in a descending order of wind velocities, the types of deformation and the values of mean velocity at the height of the top of the specimen. The practical range of mean velocity at tree-top height over which we possess ecological indicators in New England is thus seen to be from about 10 miles per hour to about 27 miles per hour. As indicators in this range we have five progressive types of deformation (brushing, flagging, throwing, clipping, and carpet) applied to a number of species from the most sensitive (white pine and hemlock) to the least sensitive (balsam and spruce).

Balsam is a good indicator in New England, throughout a range of mean annual wind velocities which includes the economically useful range. The entire scale of deformation of balsam corresponds to a remarkably short range of mean velocity at *specimen height,* running from minimal flagging at about 17 miles per hour to a 6-inch carpet at about 27 miles per hour, a range of about 10 miles per hour, for the five easily recognized stages of deformation.

It should be noted that the critical value of mean annual velocity, above which the specimen cannot grow, is not a constant. As the tree top increases in height and lengthens the distance of the circulation system joining the roots to the topmost new growth, a less severe wind regime is required to induce deformation, while, as the tree is gradually suppressed in height, progressively higher values of mean annual velocity are needed to produce deformation. Griggs says the explanation for the greater relative tenderness of tall specimens is doubtless to be sought in the little understood sap-pumping mechanism of a tree. In any event, the evidence summarized in Table IX shows that a 30-foot balsam will begin to show de-

TABLE IX. SUMMARY OF QUANTITATIVE ECOLOGY

Station	Elevation of Specimen Above Sea Level, feet	Types of Deformation of Trees — Deciduous	Types of Deformation of Trees — Coniferous Evergreens	Wind Velocity V at Height x (in Feet) of Top of Specimen above Ground, mph.	Wind Velocity at Hub Height V_{140}, mph.	Potential Annual Output, kwh./kw.
Mt. Washington (The Horn)	4100		Balsam, spruce, and fir held to 1 foot.	$V_1 = 27.0 \pm 1.0$ (Obs)	44 (Extr)	6000
Abraham	4040		Balsam, spruce, and fir held to 4 feet.	$V_4 = 21.5$ (Est)	35 (Extr)	4850
Cutts	4100		Balsam thrown.	$V_{25} = 19.2$ (Inter)	28 (Extr)	3800
Nancy Hanks	3900		Balsam strongly flagged.	$V_{30} = 18.6$ (Inter)	25 (Extr)	3300
Pico	4000		Balsam flagged	$V_{30} = 17.9$ (Inter)	21.4 (Extr)	2700
Glastenbury	3790	Birch, maple, beech, and cherry held to 40 feet partly by ice.	Balsam shows minimal flagging.	$V_{40} = 17.3$ (Inter)	19.8 (Extr)	2200
Herrick	2560	Hardwoods not held to a level.	Balsam unflagged.	$V_{40} = 15.5$ (Inter)	17.3 (Extr)	1450
Grandpa's Seward	1990 / 2100		Top killing.	$\{(V_{46} = 14.2$ (Inter) / $(V_{46} = 13.9$ (Inter)	16.7 (Inter) / 16.7 (Extr)	1200 / 1200
Biddie I	1850				15.8 (Extr)	950
Pond	1400		Hemlock and white pine show minimal flagging.	$V_{40} = 10.6$ (Obs)	12.3 (Extr)	450

formation when the mean velocity at specimen height is about 17 miles per hour, and that it will be held to a close carpet a few inches thick when the mean velocity at a height of 1 foot is about 27 miles per hour. Thus we cannot say of a specimen held, for example, to 4 feet, that the mean velocity at 4 feet is also 27 miles per hour. It is doubtless something less—how much less we do not yet know.

Nevertheless, if the ecological yardstick shown in Table IX applies generally throughout New England, then the deformation of the trees at Station X can be used directly to tell us something about the mean annual velocity at *specimen height*. And if the vertical distribution of velocity which was found to exist over Grandpa's, Pond, Seward, Biddie I, and Mt. Washington is assumed to exist also over Station X, then the mean annual velocity at hub height can be directly estimated, from the deformation at tree-top height. These assumptions require much testing before they can be accepted as the basis for quantitative estimates.

It seems clear, however, that, in interior New England at least, ecology can be made to relate sites in an order of merit; indicate which sites are of submarginal economic interest; and provide direct estimates of long-term mean annual velocity, with a degree of uncertainty which is not known today.

For example: The summit of Mt. Abraham, the southern shoulder of Lincoln Ridge, lies at 4040 feet, the same elevation as the Horn of Mt. Washington, and about 1500 feet below timber line. Healthy balsam which normally reaches a height of 30 feet in this habitat is held to a height of 4 feet on Mt. Abraham. Therefore we say that, at specimen height, the mean annual velocity is somewhat less than that which has been measured at the specimen height of 1 foot on the Horn, viz., somewhat less than about 27 miles per hour. Let us estimate this as 21.5 miles per hour, by somewhat more than splitting the difference between the mean velocity of 20 miles an hour, which causes heavy flagging of 30-foot balsams, and that of 27, which causes the 1-foot carpet.

Then, using the argument of Table VII and Fig. 41, we estimate a value for the long-term mean annual velocity at hub height of 140 feet, after the removal of the 4-foot trees, of about 35 miles per hour, with an uncertainty of perhaps 5 miles an hour either way. Entering Fig. 34 with this value for mean velocity, we estimate the long-term annual output, for a certain design, at 4850 kilowatt-hours per kilowatt. Since, in this range of mean velocity, the output is a nearly linear function of mean velocity, it follows that the uncertainty in our estimate of output is about the same as that in our estimate of mean velocity, viz., about 16 per cent either way.

In another example, 30-foot balsams are "strongly flagged." The mean velocity at tree height is then estimated directly from Table IX at 20 miles per hour. Using the same arguments as before (Table VII and Fig. 41), we estimate the mean velocity at 140 feet over the bare summit, after the removal of the trees, to be 24 miles per hour, with an uncertainty of perhaps 3 miles per hour either way.

Entering Fig. 34 with this value we find an annual output of 3150 kilowatt-hours per kilowatt, with an uncertainty of about plus 16 per cent or minus 20 per cent.

Whatever the quantitation may prove to be, the types of deformation enumerated above are extremely sensitive, and record, as do tufts on a model in a wind-tunnel, the turbulence of the air-stream over an irregular surface. Of great interest to the users of wind is the evidence that in such a terrain as high New England, the prevailing westerly winds flowing over a mountain top form standing waves of turbulence, recorded neatly and precisely by the coniferous evergreens, which indicate that *here* the output will be so much, while *there*, 250 feet away, it may be one tenth as much. Such evidence of sharp boundaries in the standing turbulence is found on the summits of Pond, Pico, Abraham, and the Horn, and along the carriage road on Mt. Washington just below the half-way house.

For example, the 4200-foot Wamsutta Ridge of Mt. Washington runs from S by E to N by W, and lies athwart the prevailing wind, which, 85 per cent of the time, has a westerly component. Wamsutta Ridge lies 2000 feet below the summit and 1300 feet below timber line. Along most of the crest of the ridge the trees grow between 15 and 30 feet high. But in one patch a hundred yards wide, the high-velocity wind-strain has been deflected downward for unknown causes, and sears the ridge. Here balsam grows only in the lee of rocks and only to a height of 1 foot. The transition from this zone of high wind to the zone of winds permitting nearly normal growth occurs in a matter of yards.

Another and a similar example occurs on Chandler Ridge, one-half mile downwind and to the east of Wamsutta Ridge, and lying 500 feet lower. Here the patch of 1-foot balsam carpet on the Horn extends a half mile from north to south, and the transition zone between the carpet and the normal 30-foot forest, while it is graduated, takes place in a matter of yards.

The ecological yardstick of long-term mean velocity is admittedly still crude, and has been developed from fragmentary data, with a scale probably applicable only to New England. But we think that, at least in interior New England, it provides a useful method of taking the first step in a wind velocity survey. It is hardly necessary to add that, before commercial development is undertaken at a potential wind-power site, the ecological survey should be followed by anemometry long enough to relate the site to the long-term mean velocity determined at a meteorological control station, as described on pp. 74–80.

Icing.

If our ecological yardstick is a crude measure of mean velocity, it is cruder yet as a measure of maximum icing, for the reason that a "candelabrum" forest may have been produced by a few exceptionally heavy ice storms unassisted by wind; or it may be the result of less severe icing, but accompanied or followed by strong winds.

Quantitation of this type of deformation seems most difficult and uncertain.

However, the negative evidence may be useful. If a mature forest shows no damage due to icing, this indicates that, during the life of these trees, maximum icing has not been adequate to break branches unaided by wind.

Summary of the Ecological Evidence

1. Occasional very severe storms do not deform trees.

2. For each species and each habitat, there is some critical minimum value of the mean wind velocity below which deformation will not occur. We do not know very much about this limit.

3. Tree deformation is a sensitive indicator of the unpredictable heterogeneity of wind-flow through and over mountains. Local transitions from prevailing very high winds (which hold the balsams to a carpet), to prevailing winds so moderate that the balsams reach normal growth without deformation occur within a matter of yards.

4. Wind-flow in mountainous country is turbulent. While the small structure of this turbulence is as yet unpredictable, it frequently consists of standing waves so nearly permanent (under the conditions of prevailing westerlies in New England) that the evidence is recorded in the deformation of the trees, particularly on down-wind slopes and shoulders, where "reverse" east-wind flagging is found, often to within 100 feet of the west-wind flagging on the summit.

5. Tree deformation is a poor yardstick of maximum icing, although absence of breakage by ice may be significant.

6. Balsam is the best indicator of mean wind velocity in mountainous New England. Deformation begins when the mean velocity at specimen height (say, 30 feet to 50 feet) reaches 17 miles per hour. The other end of the scale is reached when balsam is forced to grow like a carpet, at a height of 1 foot, by a mean velocity of about 27 miles per hour.

7. In this range of 10 miles per hour there are five easily recognized types of progressive deformation (brushing, flagging, throwing, clipping, and carpet).

8. The ecological yardstick of mean wind velocity is still of unknown accuracy, but, at least in interior New England, we believe it can be made to:

   a. Relate sites in an order of merit;

   b. Indicate which sites are of submarginal economic interest;

   c. Provide direct estimates of long-term mean annual velocity, with a degree of uncertainty which we do not know, but which may be of the order of ± 20 per cent.

9. The southern three-quarters of Lincoln Ridge appears to be the best large capacity site in Vermont. On Mt. Abraham (4040 feet), the southernmost shoulder of the ridge, the mean velocity at hub height is perhaps about 35 miles per hour and the average annual output would therefore be about 4850 kilowatt-hours per kilowatt from a 175-foot, 1500-kilowatt unit.

Chapter V

THE POWER IN THE WIND AND HOW TO FIND IT

The power in the wind can be computed using the concepts of kinetics. Power is equal to energy per unit time. The energy available is the kinetic energy of the wind. The kinetic energy of any particle is equal to one-half its mass times the square of its velocity, or $\frac{1}{2}MV^2$. The volume of air passing in unit time, through an area A, with velocity V, is AV, and its mass M is equal to its volume multiplied by its density, ρ, or, $M = \rho AV$. Substituting this value of the mass in the expression for the kinetic energy, we obtain

$$\text{Kinetic energy} = \frac{1}{2}\rho AV \cdot V^2 = \frac{1}{2}\rho AV^3.$$

If a non-dimensional proportionality constant k is introduced to convert the energy to kilowatts, then, when

$\rho = $ the density, in slugs* per cubic foot $= \dfrac{M}{g} \cdot \dfrac{1}{\text{ft}^3}$ ** ;

$A = $ the projected area swept by the turbine, in square feet $= \dfrac{\pi D^2}{4}$;

$V = $ wind velocity in miles per hour; and

$k = 2.14 \times 10^{-3}$;

the expression for the power in the wind becomes

$$\text{power in kilowatts } (kw.) = 2.14\rho AV^3 \times 10^{-3}.$$

For example, the power passing through the disc area of the test unit in a 30-mile wind with a density of 2.33×10^{-3} slugs per cubic foot is:

$$Kw. = 2.14 \times 2.33 \times \frac{\pi}{4} \times (175)^2 \times (30)^3 \times 10^{-6} = 3240 \text{ kilowatts.}$$

The Power Which Can Be Extracted from the Wind

It is obviously impossible to convert all the power of the wind into useful power. The portion that is usable is determined by aerodynamic and mechanical efficiencies (Ref. 19-B) described in Chapter X. The over-all efficiency varies with the wind velocity. In Fig. 50 there is plotted the power output at the genera-

* A slug is a unit of mass equal to the weight divided by the acceleration of gravity.
** At the test site (1990 feet), the first approximation of the 5-year mean value of the density was found, by weighting the average density for each month by the proportional output during that month, to be 0.00233 slug/ft³.

tor in kilowatts, against the wind velocity in miles per hour, for a specific turbine design rated at 1500 kilowatts and turning at 30 revolutions per minute in air with a density of 1125 grams per cubic meter, corresponding to the annual average found at 4000 feet in Vermont. This design starts generating at about 17 miles an hour, and the rate of change of output with velocity is quite steep in the velocity range 17 miles an hour to 21 miles an hour. But in the velocity range above 25 miles an hour, the relationship between output and wind velocity is very nearly linear, because of the way in which the aerodynamic, mechanical, and electrical efficiency curves are combined in this particular design. For any velocity-frequency distribution curve, there is some most economical combination of turbine speed and generator rating.

Knowing the output in kilowatts for any value of wind velocity in miles per hour, it is possible to compute the annual output at a site whose long-term velocity-frequency distribution curve is known by converting the velocities to outputs, multiplying by the hourly distribution, and summing to determine the total annual output expressed in kilowatt-hours.

A wind-turbine whose design is economical will show a maximum over-all efficiency of about 35 per cent, usually at some low value of wind velocity, say 18 miles an hour, and will convert to electrical energy about 6 per cent of the energy in the wind which annually passes through the disc area.

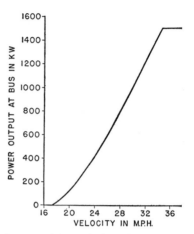

FIG. 50. The relationship between the power output at the bus and the wind velocity for a 30-rpm wind-turbine rated at 1500 kilowatts, operating at an elevation of 4000 feet in a wind-stream whose mean annual density is 1126 grams per cubic meter.

The actual monthly frequency distributions of velocity during 60 months at Mt. Washington and at Grandpa's have been converted into computed outputs. These outputs, with the output scale converted to annual output in kilowatt-hours per kilowatt, are plotted against the average velocity for each of the 120 months, in Fig. 33 (Ref. 20-B). There is considerable dispersion, as would be expected. The mean curve is also plotted. This curve refers to the test unit design, namely, a rated capacity of 1250 kilowatts, turning at 28.7 revolutions per minute, in wind whose average annual density of 1195 grams per cubic meter is characteristic of an elevation of 2000 feet in Vermont.

The variation of annual output with variation in mean annual velocity has also been computed for a unit rated at 1500 kilowatts and turning at 30 revolutions per minute, in wind whose density of 1125 grams per cubic meter corresponds to the annual average found at 4000 feet in Vermont (Fig. 34). This curve passes through eight points, determined by multiplying the eight velocity-frequency distribution curves listed in Table X and some of which are shown in Figs. 35 and 36, by the power-velocity curve of Fig. 50. The velocity-frequency distribution curves had all been smoothed, and it will be noticed that the resultant points in Fig. 34 fall on a fair S-curve similar to that of Fig. 33, but with small dispersion.

With the curve of Fig. 50, it is possible to estimate the output of this particular design at any site (in interior New England), where the mean velocity at hub height is known. A similar curve (for interior New England) could be prepared

TABLE X. SUMMARY OF THE RELATIONSHIP BETWEEN MEAN ANNUAL
VELOCITY AND UNIT OUTPUT

Station	Period	Yrs. of Obs.	Type	Assumed Speed-up Factor	Mean Annual Velocity in mi./hr.	Output in Kwh./Kw.		
						Sea Level	4000 Ft.	10,000 Ft.
1. Boston 15	1934–1938	5	Smoothed	1.00	9.37	195	169	132
2. Nantucket	1930–1934	5	Smoothed	1.00	14.95	903	807	653
3. Grandpa's	1940–1945	5	Smoothed	1.00	16.51	1263	1133	929
4. Blue Hill	1933–1939	7	Computed	1.00	18.58	1973	1801	1523
5. Blue Hill	1933–1939	7	Computed	1.20	22.28	3026	2819	2477
6. East Mt.	1931–1939	9	Computed	1.15	26.93	3809	3633	3338
7. East Mt.	1931–1939	9	Computed	1.40	33.04	4749	4564	4274
8. Mt. Washington	1933–1940	8	Smoothed	1.00	40.79	5917	5787	5558

from these same velocity-frequency distribution curves for any other design for which the power-velocity curve was known; *and the same sort of curve could be prepared for any other region for which a suitable collection of velocity-frequency distribution curves was available.*

It must be recognized that actual output over a 20-year period may depart by a few per cent from that indicated by entering the smooth curve of Fig. 34 with the value for a 20-year mean velocity, because of skewnesses in the monthly and annual velocity-frequency distribution curves. And, of course, in the case of an individual year, the departure of actual output from that indicated by entering Fig. 34 with the observed mean velocity for the year may be as much as ± 10 per cent.

The Periodic Fluctuations in Wind-Power

In Fig. 51 are plotted the computed *annual* outputs during the 5-year period, 1941–1945, at Grandpa's, Blue Hill, and Mt. Washington, each expressed as a percentage of the respective mean outputs for the five years.

Grandpa's, having the lightest of the three wind regimes, fluctuates from plus 19 per cent to minus 26 per cent, showing less stability than either the seacoast station or the high mountain station.

The Power in the Wind and How to Find It

The computed *monthly* outputs for these stations during this period are shown in Fig. 47. While the general trends are similar, some interesting comparisons are

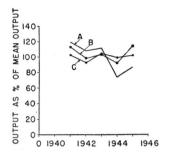

FIG. 51. Fluctuations in the annual output during a 5-year period, expressed as percentage deviations from the mean output for the five years.

Curve A   Grandpa's
Curve B   Blue Hill
Curve C   Mt. Washington

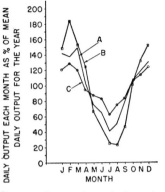

FIG. 52. Seasonal variations in output computed from five years of records.

Curve A   Grandpa's
Curve B   Blue Hill
Curve C   Mt. Washington

seen to occur in February, 1942; November, 1942; January, 1943; March, 1944; and January, 1945. They have not been investigated.

In Fig. 52 are shown the variations in computed output for the three stations throughout the year, expressed as the percentage of the mean daily output for the year, averaged over the five years. Thus at Grandpa's the daily computed output during February averaged 138 per cent of the mean daily output for the five years; and in July, 41 per cent.

It will be noted that Grandpa's *seasonal* variation in computed output stands in the middle between the more regular computed output of Mt. Washington and the less regular computed output of Blue Hill.

In Fig. 53 is shown the computed

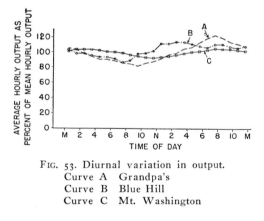

FIG. 53. Diurnal variation in output.
Curve A   Grandpa's
Curve B   Blue Hill
Curve C   Mt. Washington

mean *diurnal* variation in output at the three stations (Ref. 21-B). The least diurnal variation occurs at Mt. Washington, the most at Grandpa's, and the sea-breeze effect is noticeable at Blue Hill. At both Mt. Washington and Grandpa's, the maximum velocity occurs from 7 P.M. to 8 P.M., while the minimum occurs 2 hours later at Mt. Washington than at Grandpa's.

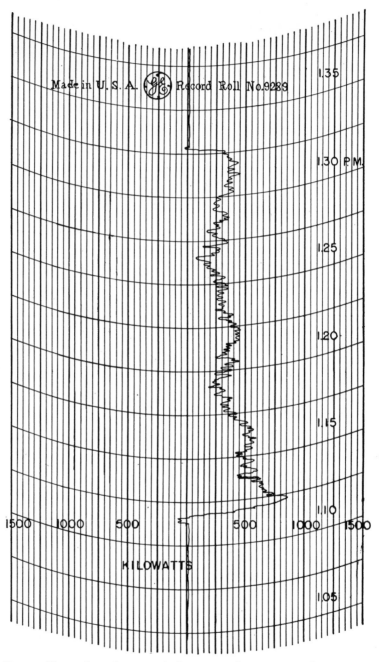

FIG. 54. Fluctuation of output during a 20-minute run on January 25, 1943, in wind varying from 19 miles per hour (output at 75 kilowatts) to 28 miles per hour (output at 870 kilowatts).

In Fig. 54 is reproduced a typical output strip-chart for a 20-minute period, showing the type of *short-term* fluctuation in output generated by the test unit when the wind velocity was varying from about 19 miles an hour (output at 75 kilowatts) to about 28 miles an hour (output at 870 kilowatts). The period covered was from 1:12 P.M. to 1:32 P.M., January 25, 1945.

Reliability of Wind-Power

The basis for study is the variation in computed aeroelectric output at Blue Hill, Grandpa's Knob, and Mt. Washington.

Figs. 55 B, C, D have been drawn to show the maximum positive and negative departures in the daily output of each month, expressed as a percentage of

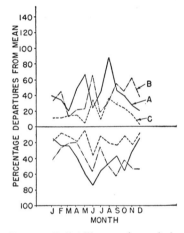

FIG. 55. Reliability of wind-power. Maximum positive and negative departures in computed monthly output from the expectancy for that month as measured over a 5-year period, for each of three stations.

    Curve A   Blue Hill
    Curve B   Grandpa's
    Curve C   Mt. Washington

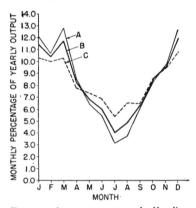

FIG. 56. Average seasonal distribution of output, computed from five years of records at two stations and interpolated for the proposed 9000-kilowatt installation at Lincoln Ridge.

    Curve A   Grandpa's
    Curve B   Lincoln Ridge
    Curve C   Mt. Washington

the mean daily output for each month during the 5-year period, 1941–1945, for each of these four stations.

It will be seen that departures from expectancy, as established by five years of experience, tend to be least at Mt. Washington, with Grandpa's tending to be more reliable than Blue Hill.

As might be expected, wind output is most reliable in the fall, spring, and winter, and least so in mid-summer.

In five years of experience at Grandpa's the computed output for one month never exceeded 163 per cent of the expectancy for that month, or fell below 39 per cent of it. The limits for an individual year were 120 per cent to 78 per cent.

For the proposed 9000-kilowatt installation on Lincoln Ridge (4000 feet), we have estimated that the reliability would be greater than at Grandpa's but less than at Mt. Washington, and that the average monthly fluctuation throughout the year would be approximately as shown in Fig. 56.

Short-Term Predictability of Wind-Power

If utility power dispatchers on an integrated system containing some fuel burning capacity knew some hours in advance what the wind was going to do, they could at times reduce some of the "floating" and "banked" standby steam capacity, thus endowing short-term predictability with economic value.

Accordingly, Willett in Cambridge attempted to predict 30 hours in advance, with confirmations 24, 18, 12, and 6 hours in advance, which one of three conditions as listed in Table XI would prevail during a 6-hour period at the test site at Grandpa's.

TABLE XI. CONDITIONS TO BE PREDICTED

No output	No winds of more than 17.5 miles per hour
Some output	
Full output	No winds of less than 34 miles per hour

The results of several hundred predictions, analyzed only qualitatively (Ref. 23-B), show that the correlation factors were high enough to be useful, running from about 0.70 for the 30-hour prediction to over 0.90 for the 6-hour prediction.

How to Select a Wind-Power Site

Aerodynamic Criteria.

After five years of increasing familiarity with the problem of site selection, we can point to no analogy between the profiles of mountains and the profiles of airfoils by which one can predict mean wind velocities at hub height within limits which will be useful.

For example, the Horn, the 4100-foot spur lying to the north of Mt. Washington (6288), has been shown in Chapter IV to be windier than the summit. *Yet directly up-wind of the Horn are two ridges, each higher than the Horn; and the lower and nearer of these ridges is well timbered on most of its crest, with evidences of far less wind in general than on the Horn.*

Another example is Mt. Ellen (4135 feet), the northern end of the 3.0-mile Lincoln Ridge. The southern end is Mt. Abraham (4040 feet). In Fig. 23 the profiles are compared. Each summit is a low nubbin sticking up above the mean height of the ridge (4000 feet). Each is a terminating shoulder, with a deep ravine to the north and south, respectively. Aerodynamically, there is little to choose. Yet the ecological evidence is very strong that the output at Abraham would be much larger than the output at Ellen.

A third example is that of Grandpa's (2000 feet) and Seward (2100 feet). They lie within 1½ miles of each other (Fig. 25) and their profiles are similar, yet the output at Seward, computed from anemometer records, is 30 per cent greater than at Grandpa's, for reasons which we do not understand.

So far as aerodynamic criteria are concerned, we must sum up by saying that we lack the basis for arranging wind-power sites even in an order of merit.

Ecological Criteria.

An ecological yardstick has been developed for use in the Green Mountains and the White Mountains of New England as described in Chapters III and IV. Where trees are deformed by wind, the long-term mean annual wind velocity at specimen height may be estimated directly by a trained observer. The criteria of tree deformation by wind are striking, and, being easily taught, are ideal as the basis for preliminary reports by hunters, timber cruisers, fire wardens, and similar persons frequenting mountain crests.

Summary of the Criteria for the Selection of a Wind-Power Site.

In sum, and repeating the lessons of Chapter IV, we believe good wind-turbine sites will be found at certain points on ridges lying athwart, or nearly athwart, the flow of the prevailing wind; that they are not likely to be found among up-wind foothills, although they may be found on down-wind shoulders; and that the deformation of the trees will serve to arrange a list of potential sites in an order of merit and will indicate whether the wind-flow at a potential site is sufficiently interesting to warrant a program of anemometry.

Specification for a Regional Wind-Power Survey

Case 1. The region is timbered. Until more field work is carried out, there is no reason to believe that the ecological yardstick developed in the Green Mountains and the White Mountains of New England, and described in Chapters III and IV, would have the same scale values in other regions. However, there is no reason to doubt that ecological criteria would permit arranging windy sites in any timbered area in an order of merit, and therefore it would seem logical that the first steps in any regional survey of wind velocity in timbered regions should consist of field trips by an ecologist.

If a meteorological control station, with anemometer records for at least a 10-year period, does not exist within 50 miles in distance and within 2000 feet in elevation, of the sites to be surveyed, then it will be necessary to create such a control station, using the ratioing method described in Chapter IV.

Accordingly, we would suggest the following steps for a regional wind-power survey in timbered country:

1. An illustrated questionnaire directed to hunters, timber cruisers, fire wardens, high trappers, hikers, and others, to bring out ecological evidence of local

high winds in regions thought to be generally windy, and in which occur ridges lying athwart the prevailing wind direction. The illustrations should cover brushing, throwing, flagging, clipping, and carpets.

2. Field trip by an ecologist accompanied by a meteorologist, to evaluate the evidence collected by the questionnaire, and to arrange potential sites in an order of merit.

3. If necessary, the creation of a mountain-top anemometer control station, the wind velocity at which is ratioed into the long-term mean velocity of the local free-air, the latter being determined in turn by ratioing the local radar pilot-balloon runs to data from the nearest upper-air station, whose 20-year average velocity is taken as the base.

The exposure of the anemometer is everything. Exposures down-wind of even such open structures as the Christmas Tree result in errors of 5 per cent over a long term. An anemometer should be mounted at mast head, well in the clear, and any lower anemometer should be mounted on a strut at least 10 mast-diameters in length, and pointing up into the prevailing wind, at 45 degrees to the prevailing direction.

4. The long-term mean velocity, at observation stations lying within 50 miles in radius and 2000 feet in elevation of the control station, is determined by ratioing short records into the record at the control station, using selected data, as described in Chapter IV.

5. If this program is made to include the "mean" periods, usually April–May or September–October, uncertainties in ratioing the observational data to the control data will be held to a minimum, and 90 days of selected record, in conjunction with the ecological evidence, should provide estimates close enough ($\pm$ 20 per cent) to guide policy, since in the range of mean velocity at hub height (150 feet), that is, from 25 miles an hour to 45 miles an hour, output is a nearly linear function of mean velocity, and the uncertainty in the predicted output would, therefore, be about the same as the uncertainty in the predicted mean velocity (Fig. 33).

6. Once the wind regime for the region is established on a quantitative basis, the ecological yardstick, which heretofore has been lacking, can be supplied with confidence, and preliminary estimates of output at many timbered sites can be made directly, without anemometry.

Case 2. The region is not timbered. Where the region is not timbered it will be necessary to set up many more observation stations. As explained in the previous section, we are not yet able to study a range of mountains and, based on geometry alone, arrange a group of potential sites even in an order of merit. For this reason, a wind velocity survey in an untimbered region would consist solely of Steps 3, 4, and 5.

The Power in the Wind and How to Find It

Summary

1. The power in the wind in kilowatts may be given by the formula

$$\text{Power} = \text{Kw.} = 2.14\rho A V^3 \times 10^{-3}$$

where

ρ = the density, in slugs per cubic foot $= \dfrac{M}{g} \cdot \dfrac{1}{\text{ft}^3}$;

A = the projected area swept by the turbine, in square feet $= \dfrac{\pi D^2}{4}$;

V = wind velocity, in miles per hour.

2. *Estimating annual output.* A curve has been developed which gives the annual output (in interior New England), for a specific design, when the mean annual velocity is known. The method may be applied to any region for which velocity data are available.

3. *Criteria of site selection.* The best criterion for a survey of wind velocities in windy forested country, preliminary to anemometer measurements, is the deformation of coniferous evergreens.

4. *Specifications are suggested for a regional wind-power survey,* based on experience summarized in Chapter III.

Chapter VI

DESIGNS OF OTHER BIG WINDMILLS, 1920–1933

Toward the close of the First World War, and immediately following it, scientists in France, Germany, and Russia became interested in developing a modern theory of the windmill, in conjunction with war-born progress in the theory of the air-screw. Joukowsky, Drzewiecki, Krassovsky, and Sabinin in Russia, Prandtl and Betz in Germany, and Constantin and Eiffel in France, were the architects of modern windmill theory. Betz was the first to show that no windmill could extract more than 16/27 (about 59.3 per cent) of the energy passing through the area swept.

The war had stimulated developments, not only in propeller design, but also in light metals and in radio. Demobilized youngsters, with Yankee ingenuity, mounted the propellers from their Curtiss Jennys on the tops of barns to drive small battery-charging sets for their homemade radios; and engineers were active in designing larger windmills, in accordance with modern theory.

It is not a purpose of this chapter to attempt a definitive history of the development of the theory and practice of windmill design. In the Bibliography there is a fairly complete list of source material for such a history. Rather, I shall review the principal designs of the largest windmills, and indicate why no one of them appeared quite satisfactory.

Three paths were followed by the pioneers. A group, including Flettner and Mádaras, sought to harness the Magnus effect, the thrust exerted by a cylinder spinning in a wind-stream (Fig. 57). Flettner had crossed the ocean in the rotor ship *Baden-Baden,* propelled by this thrust. Another path was explored by Savonius, in Finland, who built S-shaped rotors, with the axis vertical. A third group, Kumme and Bilau in Germany, Darrieus in France, the staff of the Central Wind Energy Institute in Moscow, Fales in the United States, and others, had evolved designs, some small, some large, which were variations of the high-speed propeller type.

It is convenient to distinguish between windmill designs by means of the tip-speed ratio—the ratio between the peripheral speed of the tip and the true wind velocity. Dutch windmills possessed a low tip-speed ratio, of 1.0 or less. "American" multi-bladed windmills had a higher ratio, of about 2.0. Modern windmills of the propeller type were high-speed, with working ratios of 5.0 and more. A low tip-speed ratio is associated with low efficiency but with the ability to start under load. A higher ratio means higher efficiency, but inability to start under load.

Flettner in 1926 built a windmill (Ref. 4-A), the four blades of which were tapered rotating cylinders, driven by electric motors. With a diameter of 65 feet 8 inches, it was rated at about 30 kilowatts in a wind of 23 miles per hour. Each

FIG. 57. Flettner's rotor ship "Baden Baden" which successfully crossed the Atlantic in 1925.

cylinder was 16 feet 5 inches long, 35.5 inches in diameter at the outer end, and 27.5 inches at the inner end. The tower was 108 feet high. The windmill drove a direct current generator through a 1:100 transmission.

The disadvantages of this design seemed real enough to warrant discarding it on von Kármán's judgment, without a rigorous analysis. Von Kármán pointed out that a rotor-type turbine is considerably less efficient, per unit area swept, than the propeller-type, without compensating savings in weight; that regulation to constant torque, by control of the speed of spinning, would be made expensive by the momentum of the spinning masses; and that, finally, the higher thrust would require greater tower investment per kilowatt. Thus there seemed no inducement to explore the Flettner design further.

Mádaras proposed harnessing the Magnus effect in a different way (Ref. 5-A). On a circular track he would operate a train of flat cars. On each car would be a spinning cylinder, 90 feet high by 28 feet in diameter, driven by an electric motor. A component of the Magnus thrust would lie parallel to the track, and would propel the train. The component of the thrust at right angles to the track would be absorbed in the heat of friction at the contact between the flange and the rail. Elec-

FIG. 58. The test of the pilot model of the Mádaras rotor at Burlington, New Jersey, October 1933.

tric generators in the car axles would generate power, transmitted by the rails to a switchboard for distribution. The cylinders would reverse rotation twice in each circuit of the track.

The system had a high total of aerodynamic, mechanical, and electrical losses and did not lend itself to the mountain-top locations where the wind was to be found. Further, its cylinders for extracting energy from the wind were, of necessity, mounted close to the ground, in the zone of frictional retardation, where the wind velocity is low.

A single full-scale cylinder was constructed on a concrete base at Burlington, N. J. (Fig. 58) and put under test in October, 1933. The Magnus thrust measured when the cylinder was spun in a light wind conformed to previous experience, but did not provide the economic basis for successful competition with other forms of power generation.

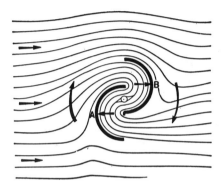

FIG. 59. Horizontal cross-section of wing rotor. Note that "A" is propelled by wind spilling from "B".

The *Savonius* rotor (Ref. 6-A) was a vertical cylinder sliced in half from top to bottom, the two halves being pulled apart by about 20 per cent of the diameter (Fig. 59). In principle it resembled a cup anemometer, with some recirculation of flow which in effect converted that cup which was coming backwards up-wind, into the second stage of a two-stage turbine.

The Savonius design possessed fairly high efficiency (31 per cent), but was inefficient per unit of weight, since all the area swept was occupied by metal (Fig. 60). Thus, in order to develop 1000 kilowatts in a wind of 30 miles per hour, a cylinder about 100 feet in diameter by 300 feet in height would be required, presenting 30,000 square feet of metal to do work which could be done by a two-bladed 175-foot turbine with about 1/30th as much metal area.

The *Kumme* design (1920) (Ref. 7-A) (Fig. 61) consisted of six blades, with rather complex guying, and was presumably operated at a moderate tip-speed ratio. The vertical generator on the ground was an attractive idea, which, however, did not stand up under economic analysis. When adequate provision was made to absorb the forces which cause the windmill to "walk" around the horizontal bevel gear aloft, throwing it out of yaw; and to keep the long flexible shaft aligned, it was found that the generator on the ground was more costly than aloft.

The *Darrieus* design was one of great refinement and elegance. In 1929 the Compagnie Electro-Mécanique erected at Bourget, France, a two-bladed unit, 20 meters in diameter (Ref. 8-A) (Fig. 62). The operating tip-speed ratio was very high (10.0), probably higher than could be justified economically.

FIG. 60. The Savonius wing rotor arranged for pumping water.

It can be seen that coning was frozen, giving a built-in sweep-back, which reduced the operating bending moments at the blade roots. Since the shaft was hori-

FIG. 61. The Kumme wind-turbine built in Germany in 1920.

zontal, sweep-back also provided stability in yaw, permitting the elimination of a tail vane or a yaw mechanism.

For a small windmill, without great differences in velocity at different points throughout the disc area, sweep-back seems a sensible way to reduce the average

FIG. 62. The 20-meter Darrieux wind-turbine erected at Bourget in France in 1929.

working stress at the blade root. But in a disc area so large that steep gust gradients are the rule, I felt that coning would be better. As will be told, in the test unit we began operating with the coning strongly restrained, but we found that the freer the coning the smoother the operation.

Sweep-back will not, of course, give stability in yaw when the axis of rotation is inclined to the horizontal. But a horizontal shaft incurs more debits than credits.

The *Russian* design was bold and practical, if one has in mind the limitations under which they were then working. In May, 1931, after two years of wind measurement, a wind-turbine 100 feet in diameter was put in operation on a bluff near Yalta, overlooking the Black Sea, driving a 100-kilowatt, 220-volt induction generator, tied in by a 6300-volt line to the 20,000-kilowatt, peat-burning steam-station at Sevastopol, 20 miles distant (Fig. 63) (Ref. 9-A).

Regulation was by pitch control. Pitching moments were brought nearly to balance by adjustable counterweights on struts, and pitch was varied as a function of rotational speed, by means of centrifugal controls acting on the offset ailerons. Fluctuation in output in a 7-minute run ranged from about plus 20 per cent to minus 15 per cent.

Thrust was taken up by an inclined strut whose heel rested on a circular track on the ground. Automatic control of yaw was obtained by driving this strut around the track, by means of a 1.1-kilowatt motor responsive to a wind-direction vane aloft. Hub height was about 100 feet.

The generator and controls were aloft in a streamlined house. The axis of rotation was inclined at 12 degrees to the horizontal, but in the sense that the up-wind end was high. The blades were up-wind of the tower. The main gears were of wood. The blade skins were of roofing metal. The maximum aerodynamic efficiency at 30 revolutions per minute was 24 per cent, reached at a tip-speed ratio of 4.75. The wind velocity at which rated output was reached was 24.6 miles per hour.

In a wind regime characterized by a mean annual velocity of 15.0 miles per hour, the annual output reported was 279,000 kilowatt-hours, generated at an average level of 48.4 kilowatts in the windy month of March; 18.0 kilowatts in the quiet month of August; and 32.0 kilowatts for the year.

After two years of experimental operation, it was planned in 1933 to add two units of 100 kilowatts each, but of varying design, as steps intermediate to the ultimate installation of the 5000-kilowatt units called for by the Second Five-Year Plan.

I felt that the principle weaknesses of the Russian design—low efficiency, crude regulation and yaw control, high weight per kilowatt, and induction generation— had been imposed upon the designers by the state of industry in Russia, where heavy forgings, large gears, and precision instruments were unavailable.

Another proposal considered was that of *Fales,* who had been among the first

FIG. 63. The 100-foot, 100-kilowatt DC Russian wind-turbine erected near Yalta on the Black Sea in 1931.

to become interested in high-speed windmills. He proposed a single high-speed blade, suitably counterweighted. Among the objections to this was the compelling one that, under icing conditions, the counterweight could no longer be relied upon to counterbalance the iced-up blade.

FIG. 64. Design proposed by Honnef of Berlin in 1933. This wind-turbine was to stand 1000 feet high and the inventor rated it at 50,000 kilowatts.

Honnef of Berlin suggested a bold design in 1933, a model of which is shown in Fig. 64 (Ref. 10-A).

Impressed by the increase of wind velocity with height above the level ground, he proposed that the tower be "about 1000 feet in height, or even higher." He shows a well-designed tower, which is particularly strong in torsion. In an effort to reduce the tower cost per kilowatt, Honnef designed it to support five 250-foot turbines, in a light framework free to yaw, and supported as a unit on bearings at the head of the tower. Gears are a costly bottleneck in any design of a large windmill. Honnef solved this problem by eliminating it. His generators were built into his turbines, each of which was double, consisting of two counter-rotating members. At the rim of each of the two turbines in a pair were located suitable copper and iron, such that one member of the pair acted as the rotor and one as the stator in a direct current generator. The two speeds were 10 and 17 revolutions per minute, giving an equivalent generator speed of 27 revolutions per minute.

I felt that Honnef had exaggerated the importance of height. Knowing the rate at which wind velocity increases with height and also the rate at which tower cost increases with height, it is possible to plot, for any disc area, the increase in annual output in kilowatt hours with increasing tower height; and, likewise, the increase in the total investment of the wind generating station as tower height increases. In collaboration with the American Bridge Company, and my meteorological colleagues, I had studied this relationship for a variety of tower designs and turbine sizes, and had concluded that, for a 200-foot diameter turbine on a ridge, there was no justification for a tower higher than 150 feet. It did not seem likely that the most economical height over level ground would be much greater.

Further, the Honnef scheme of a panel of five 250-foot turbines presented such formidable problems in stress analysis that it had no place in any first attempt, which should be limited to one large turbine.

Finally, I could find no one who could think of an economical structure whereby the air-gaps in the five 250-foot generators could be so maintained as to yield reasonably high electrical efficiencies when such a structure was exposed to the heterogeneous and violent stresses imposed by wind, sun, and ice.

Many other designs were considered, including several variations of the attempt to extract energy from cross-sections of the wind-stream larger than the diameter of the turbine disc. This proposal is an enticing will-o'-the-wisp, which so far has not been realized.

SUMMARY

A review of the prior art led me to propose a design of my own. S. Morgan Smith Company considered the prior art, and adopted my design as the basis for engineering studies of what became known as the Smith-Putnam Wind-Turbine.

Chapter VII

DEVELOPMENT, FABRICATION, AND ERECTION OF THE TEST UNIT, 1939–1941

Selection of Basic Design

No one of the published designs of the larger wind-turbines, described in the previous chapter, appeared attractive economically.

It seemed to me that the most economical design would be a larger wind-turbine than any yet built, with two, or possibly three, propeller-type blades, having a radius about three-quarters the height of the supporting tower, operating at a high tip-speed ratio, and located on a ridge which would speed up the "free-air" velocity of the wind.

"Three-quarters" was a rough number, which merely illustrated my conviction that, if it was unprofitable to extract energy from the frictionally-retarded winds close to the ground, it was equally uneconomical to debit the output of a small turbine with the cost of a tall tower underneath it.

Of course I had no notion of the most economical proportions of the turbine, or the proper shape of the ridge. Before the first of these two problems could be investigated, it was necessary to clarify the basic design. I will not follow all the threads of the development of the main features of the design, which can be summarized as follows:

1. *Ability to Spill the Power from High Winds.*

A wind-turbine to produce 1000 kilowatts from a 30-mile-an-hour wind of sea-level density, and operating with an over-all efficiency of 30 per cent at its rated capacity, must have a diameter of 175 feet (Chapter V). But this great structure must be gale-proof, a need which implies the ability to reduce the turbine area somehow, when a gale impends. To accomplish this reduction one could allow the turbine disc to swing out around the vertical axis until the disc lay parallel to the wind (Fig. 65A); or one could accomplish the same result about the horizontal axis (Fig. 65B); or one could feather the blades to a position of low lift and minimum drag. The latter seemed to be the most attractive solution mechanically, and the gustiness of the wind made it preferable aerodynamically. Furthermore, the mechanism for feathering the blades would be merely an extension of the means for controlling the pitch; and control of the pitch had been selected as the method of regulation.

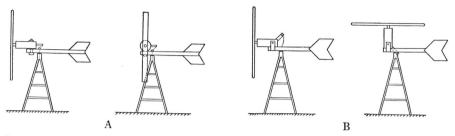

FIG. 65. A conventional means for spilling power in high winds. (A) Schematic device for rotating the turbine out of the wind about the vertical axis. (B) Schematic device for rotating the turbine out of the wind about the horizontal axis.

2. *Regulation of Torque Input.*

Hamilton Standard Propellers and others had developed controllable pitch to the point where it was accurate and dependable. Kaplan in Czechoslovakia had introduced the principle to the hydraulic turbine known in this country as the

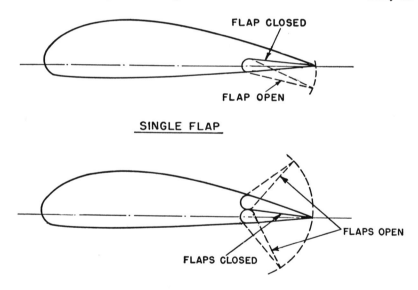

FIG. 66. Flaps, for the purpose of controlling torque.

Smith-Kaplan turbine. In each case the pitch was controlled by the action of a centrifugal fly-ball speed governor.

In the hands of the Woodward Governor Company, this type of control had been developed until it would respond to a speed change of as little as 0.01 of 1 per cent.

Other methods of achieving regulation were possible. Instead of altering input torque by altering the pitch of the blades, one could alter it by flaps (Fig. 66), or by spoilers. Any one of these three means of controlling input torque could be controlled by a torque-meter rather than by a speed governor.

Although von Kármán pointed out some of the advantages of flap control over pitch control, less was known about it, and Frank Caldwell, Chief Engineer of Hamilton Standard Propellers, and the Engineers of the Budd Company recommended the adoption of pitch control, at least for the early test units, since it did not then seem likely that the economic differences between these two controls would make or break the project.* This argument applied even more forcefully in the case of the torque meter, which would require a heavy development program before we could feel the same confidence in it that we felt in a Woodward Governor.

So, I decided to specify the regulation of input-torque by control of pitch through a Woodward fly-ball speed governor.

It was clear that regulation by any means would be assisted by a high inertia in the turbine—that is, the power output of a large heavy turbine would be smoother than that of a small light turbine. So, the larger the most economical unit proved to be, the better, so far as regulation was concerned.

3. *Number of Blades.*

It was found that the increase in annual output from the addition of a third blade would be about 2 per cent, based on estimates by von Kármán (Ref. 3-B), and would not justify the increase in the investment. Accordingly, the two-blade design was adopted.

4. *Coning.*

To reduce the heavy bending moments in the root sections of the blades, I proposed to allow the blades to cone.

By coning is meant the freedom of each blade, independently of the other, to move down-wind (positive coning) or up-wind (negative coning), through the angle illustrated schematically in Fig. 67. The instantaneous value of this angle is the result of a balance between the wind pressure on the blades, the centrifugal force, the angular momentum of coning, and any coning-damping that may be applied.

The idea was not new. Autogyro rotor blades were free to cone about a vertical axis. Propellers had been built free to cone a few degrees, with a coning angle that was forward. Later, propellers and wind-turbines were designed without freedom to cone, but with a built-in coning angle, in order to keep the bending moments to a minimum under working conditions. But built-in coning did not seem adequate when dealing with diameters of 175 feet, with each blade being loaded by the local

* But see Chapter XII.

wind with which it was immediately in contact. In order to allow my blades to cone, it was necessary to mount them down-wind of the tower. This provided an added advantage, by permitting inclination of the axis of rotation.

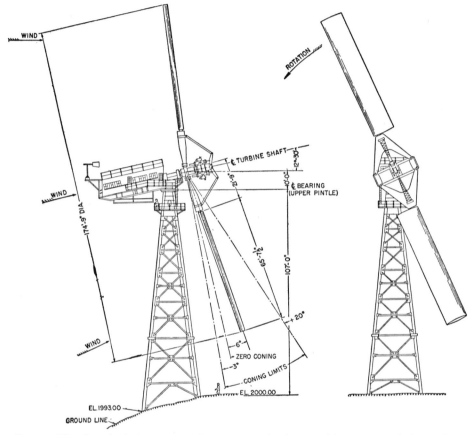

FIG. 67. The Grandpa's Knob 1250-kilowatt test unit of the Smith-Putnam Wind Turbine.

5. *Inclination of the Axis of Rotation.*

As previously explained, I counted on finding in New England, lying athwart the prevailing westerly winds, a ridge whose aerodynamic characteristics would tend to speed up the free flow of the wind. The streamlines would be most compressed just over the crest of the ridge, so the turbine disc would be located there (Fig. 12). But at this point the streamlines would lie at some angle to the horizontal, say 10 degrees or 20 degrees (Fig. 67). Therefore, it would be best if the turbine disc could somehow be tilted out of the vertical in order to lie more nearly at right angles to the streamlines.

Mounting the blades down-wind permitted tilting the axis of rotation at will. Nothing was known about the best angle of tilt. In the first place, I did not know the average angle of inclination of the streamlines. In the second, I did not know how the output from the disc area would vary as the plane of the disc was tipped out of the wind. I assumed that on any ridge whose profile was not so abrupt as to cause severe rupture of the flow (Fig. 11), the streamline angle would be less than 20 degrees. Von Kármán assured me that the output would not be sensibly diminished by tipping the plane of the turbine disc out of the wind through a small angle (Ref. 3-B).

There was a further consideration. A main shaft with a long overhang would cause high bending moments. Minimum overhang was desirable. A study was made of tower shapes to find the most economical slope of the tower legs. The value of 8 degrees from the vertical was determined by Wilbur. Therefore, an angle of inclination of the axis of 8 degrees, with minimum overhang of the shaft, would cause the blade tips to pass close to the tower legs. To avoid this close passage and to provide for some negative coning, I chose the value of $12\frac{1}{2}$ degrees for the inclination to the horizontal of the axis of rotation, feeling that this would lie within $\pm$ 5 degrees of the prevailing mean flow, and would provide adequate clearance for the blade tips (Fig. 67).

6. Yaw.

Small windmills are usually faced into the wind by a tail vane. The large Dutch windmills were "walked" into the wind by a beam on a wheel rolling along a circular path on the ground. The Russians copied this idea, adding an electric motor to drive the heel of the beam around a mono-rail track.

A tail vane adequate to control a large wind-turbine in yaw must possess a great area (or the equivalent) on an arm of some length. It seemed simpler to cause the turbine to follow in yaw the indications of a yaw vane (Fig. 67), by means of suitable servo-mechanisms and a yaw motor capable of driving the turbine around a bull-gear at the head of the tower.

7. Location of the Generator.

If the generator could be located on the ground, and mounted vertically at the foot of a vertical shaft driven by a system of bevel gears aloft, it would be easier to maintain, and weight aloft might be reduced. But studies of such a layout, in comparison with one in which the generator was aloft, showed an advantage for the latter design of several dollars per kilowatt. Accordingly I decided to specify that the generator be mounted aloft.

8. Type of Generator.

Knight told me that most utility customers would prefer a synchronous generator, and that, if an induction generator was supplied, it would usually be necessary to back it up with some condenser capacity. Studies were made of these two generating systems, and it was found that there was little or no economic advantage in induction generation. Since a system of regulation which proved adequate for synchronous generation would also be adequate for induction generation, and since the latter seemed less generally desirable, a synchronous generator was specified.

The economics of direct-current generation, with subsequent conversion to alternating current, were prohibitive in 1934, and remain so in 1946.

9. *Coupling.*

Once a synchronous generator is phased-in to a high-line, it will tend to stay in synchronism with the alternating current flowing through the high-line. To force it out of phase requires the application of a good deal of torque—perhaps 2 or 3 times the rated torque—such as might occur if the governor should fail.

But a centrifugal speed governor acting against such a violently surging "head" as a wind-stream, requires a generous range of speed-change so long as the generator remains phased-in. To provide a greater speed-change, or slip, we inserted between the generator and the gear a coupling capable of slipping (Fig. 67).

Two types of coupling were available—the electric induction type and the hydraulic type. The hydraulic coupling had been used in many applications. The electric coupling was of recent development and had been used principally in marine service at rather low speeds. In addition, the electric coupling was heavier and more expensive than the hydraulic coupling, which was accordingly specified for the test unit.

The electric coupling proposed for the production unit is a true torque-limiting device, which will limit input torque to the generator to not more than 110 per cent of rated torque.

10. *Pintle Shaft and Bearings.*

The simplest solution for the pintle shaft and bearings seemed to be a large vertical pipe supporting the horizontally rotatable (yawable) structure aloft, and in turn supported by radial and thrust bearings in the upper tower structure.

11. *Economics of a Battery of Units.*

An isolated unit would require its own road for erection, its own switchgear, and its own high-line. To reduce these unit costs, a battery of 5 or 10 units was assumed to be tied together at a common switchboard.

The Best Proportions for a Wind-Turbine

Having settled on these design features and being encouraged by Knight, I began a systematic search for the most economical size. With Knight's assistance, prices were obtained on some seventy-odd generator-gear combinations in a wide range of capacities and speeds; and from the Bethlehem Steel Corporation smooth curves were obtained from which it was possible to estimate weights and costs of four-legged towers in a wide range of heights and loadings. Rossby advanced a hypothetical vertical distribution of wind velocity over a crest, and I assumed a hypothetical velocity-frequency curve, with a mean annual velocity of 25.0 miles an hour.

The computations are described in Chapter IX and are summarized graphically in Fig. 68, which shows that the most economical unit is, apparently, rated at

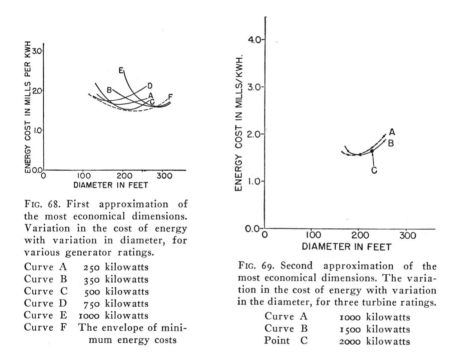

FIG. 68. First approximation of the most economical dimensions. Variation in the cost of energy with variation in diameter, for various generator ratings.

Curve A   250 kilowatts
Curve B   350 kilowatts
Curve C   500 kilowatts
Curve D   750 kilowatts
Curve E   1000 kilowatts
Curve F   The envelope of minimum energy costs

FIG. 69. Second approximation of the most economical dimensions. The variation in the cost of energy with variation in the diameter, for three turbine ratings.

Curve A   1000 kilowatts
Curve B   1500 kilowatts
Point C   2000 kilowatts

600 kilowatts, driven by a turbine with a diameter of 226 feet, and able to produce energy at the switchboard at the foot of the tower at a cost of $0.0015 per kilowatt-hour (Ref. 10-B). When talking with others, I first added 0.50 mil to this estimate; later 1.0 mil, bringing the earliest public estimate to $0.0025 (2.5 mils) per kilowatt-hour.

As described in Chapter IX, one of the first steps taken by the S. Morgan Smith Company was to review this estimate. The work was carried out early in 1940 at the Budd Company plant (Ref. 24-B), and led to the conclusion that the best unit would have a diameter lying between 200 feet and 235 feet, a capacity lying between 1500 kilowatts and 2500 kilowatts, a generator speed of 600 revolutions per minute, and blade type 2-M, as shown in Fig. 69.

Such a unit, in the wind regime then assumed by Petterssen to have a mean annual velocity of 28.8 miles an hour, would show an energy cost at the switchboard at the foot of the tower of $0.0016 (1.6 mils) per kilowatt-hour.

Wilbur found that the best proportions would not vary greatly within the limits of most wind regimes which would be commercially interesting. This point has not been thoroughly investigated, but in studying a large number of wind regimes, we have found no reason to doubt that it is an acceptable assumption.

Proportions Selected for the Test Unit

When it was found that the results of our second study confirmed in a general way the results of my first approximation, the S. Morgan Smith Company, on March 10, 1940, decided to move ahead with the project. The question of a small test unit, 25 feet or so in diameter, was explored. It was felt that the secret of smooth regulation lay in high inertia, which might not be provided by a small unit; the design, fabrication, and testing of which would in any case cost nearly as much as would the full-scale unit. To eliminate the risk of poor regulation and other scale effects, the full-scale test unit was decided upon, in the smallest size and on the shortest tower thought to be characteristic of the range of economical sizes. Accordingly, S. Morgan Smith Company selected a rating of 1250 kilowatts, a diameter of 175 feet, a generator speed of 600 revolutions per minute and a hub height of 125 feet, as the proportions for the smallest test unit which would be representative of the most economical size.

As soon as George Jessop had made this decision, the engineering of the test unit began. At first this work was carried out informally by Dr. Wilbur, under my nominal supervision in my capacity as Project Manager, but, in June, on my recommendation, Dr. Wilbur was formally made Chief Engineer of the Project, and this phase of the work thereafter came under his direction.

Description of the Test Unit

The test unit was a two-bladed wind-turbine, connected through speed-increasing gears and a hydraulic coupling, to a synchronous generator, the whole rotatable on top of a 110-foot tower. Fig. 67 shows the schematic layout of the turbine, with the principal dimensions (Ref. 25-B).

The blades were of constant chord and the cross-section corresponded to the airfoil designated as N.A.C.A. 4418. The blades were supported in the A-frames by shafts, commonly known as the blade shanks, free to rotate in bearings, under the control of the pitch controlling mechanism (Figs. 67 and 1). This mechanism consisted of a hydraulic cylinder connected, through links, cranks, and torque tubes, to the blade shanks (Fig. 70). The A-frames were free to rotate over a limited amount of travel. This movement of the A-frames and the blades, in a plane parallel to the wind direction, is known as coning (Figs. 67 and 1). The limit stops of this motion were spring cushioned and combined with a viscous damping mechanism. The connection between this mechanism and the A-frames was made by means of rocker arms and coning links (Fig. 1). The supporting structure for these damping mechanisms, and also the pitching cylinder, was known as the tailpiece. The tailpiece and A-frames were all connected to the hub-post. The hub-post in turn was connected to the main shaft by means of a taper fit and wedge keys.

The main shaft was a hollow steel forging carried on anti-friction bearings in large cast-steel housings. The center of the shaft contained oil tubes to carry oil

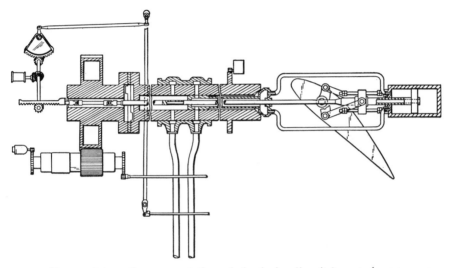

FIG. 70. Schematic representation of the hydraulic pitch-control system.

to both sides of the pitching cylinder in the tailpiece (Fig. 70), and wires to the instruments which were used to measure coning angle and coning-damping pressures. There were slip rings on the shaft for the electrical connections and an oil head for the hydraulic connections. The connection between the main shaft and the speed-increaser gear was by means of an Oldham-type coupling. The speed-increaser gear was of the divided-load type, having two jack shafts. The first step-up had spur gears and the second, continuous-tooth herringbone gears. The single high-speed output shaft was connected to the hydraulic coupling through a Fast flexible coupling. On this high-speed shaft close to the gear box was placed a gear arranged to be driven either by an electric motor or by hand, for positioning of the blades for maintenance purposes. This mechanism, which was disengageable, also served as a lock to hold the blades in position. The hydraulic coupling was overhung on the generator shaft and coupled directly to it (Fig. 1). The hydraulic coupling was a 48-inch traction-type coupling manufactured by the American Blower Company.

The generator was a General Electric synchronous machine, rated at 1250 kilovolt-amperes at 2400 volts, operating at 600 revolutions per minute, and with a direct connected exciter.

The entire turbine mechanism was carried on a stiff box-type girder called the pintle girder, which was about 40 feet long by 8 feet wide by 5 feet deep (Fig. 1). A vertical pintle shaft rigidly fixed to this girder was supported by anti-friction bearings in the tower cap (Fig. 1). This cap was a weldment attached to the top of the tower, which was of bolted construction.

Attached to this tower cap was a horizontal bull gear 10 feet in diameter (Fig. 1). A mechanism consisting of a hydraulic motor and a speed-reducing gear drove a pinion meshing with this bull gear and rotated the entire assembly aloft above the axis of the tower. This yaw-mechanism was under the control of a damped yaw vane (Fig. 67) which served to keep the turbine always in proper relationship to the direction of the wind.

Also mounted on the pintle girder were the speed-sensitive governor and its auxiliary equipment. This auxiliary equipment consisted of an oil pump mechanically driven by the turbine, an oil pump driven by an electric motor, a sump tank, an accumulator tank, and the necessary piping and controls.

Also mounted aloft were the electrical controls for space and oil heaters, manual controls for the yaw-mechanism, safety devices, and emergency shut-down switches.

The electrical connections were carried down the hollow pintle shaft to a series of slip rings where they were transferred to the stationary structure. The switchgear was housed in a concrete building some 300 feet from the foot of the tower, and partially protected from flying ice by a shoulder of the summit. The transformers were installed in the open alongside this building where the high-line terminated.

The test unit of the Smith-Putnam Wind-Turbine was designed to be a wholly automatic unattended station. Automatic control operations, shown schematically in Fig. 71, were made dependent on wind velocity, as indicated by the action of the propeller under four general conditions. These conditions were:

1. Wind velocity below that necessary to generate power, that is, less than V_{on}.
2. Wind velocity above V_{on} and less than that necessary to give rated power (V_{rated}).
3. Wind velocity above V_{rated} and less than the maximum allowable operating velocity (V_{off}).
4. Wind higher than V_{off}.

In the first condition, the turbine blades were set at a predetermined angle to give not more than 14 revolutions per minute. A speed-sensitive relay closed after a time delay and initiated the starting sequence. The blades were moved to design blade angle and the unit picked up speed. The speed was controlled by the gov-

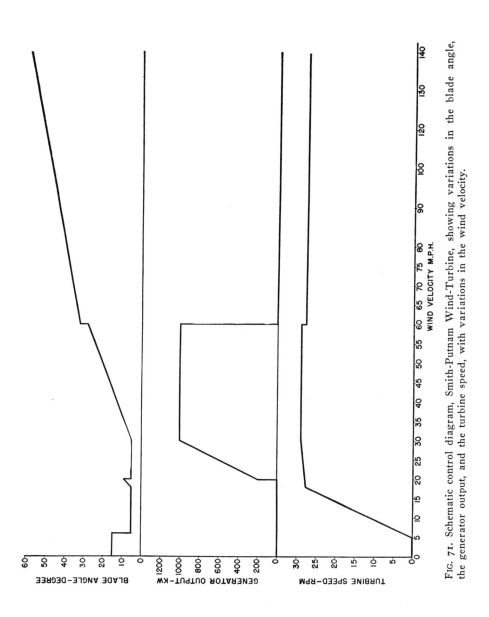

FIG. 71. Schematic control diagram, Smith-Putnam Wind-Turbine, showing variations in the blade angle, the generator output, and the turbine speed, with variations in the wind velocity.

ernor at speed-no-load. As the velocity increased the governor adjusted the blade angle to keep the speed constant. When the blade angle reached 9 degrees there was sufficient wind to load the generator to about 200 kilowatts. At this point the synchronizing cycle was initiated. Briefly this cycle was as follows: speed matching by control of governor speed adjustment; closing of line breaker; and application of field excitation.

The output in the range V_{on} to V_{rated} was entirely dependent on the wind velocity. Above V_{rated} the propeller speed, and therefore the output, was held nearly constant by the governor. At first it was thought it would be necessary to take the turbine off the line in winds above 60 miles an hour, because of the greater energy in the gusts and the difficulties of regulation. In this condition the generator was disconnected from the line and the turbine allowed to idle at some predetermined speed (Fig. 72).

The above description is by no means complete. Most of the sequences included time delays, interlocks, and various other devices (Ref. 26-B).

Some Design Problems

In completing the design of this test unit, Wilbur was faced with several design problems. His major problem was how to complete the design at all, when the various designing groups were located at California Institute of Technology, in Cleveland, York, Philadelphia, Boston, and other places; and when the design consultants were scattered from coast to coast and partially occupied with other affairs. The onrushing war created a compelling pressure, in the face of which it had been decided to order major forgings in May, 1940, before the design of the structure, or even its stress analysis, had been completed.

This calculated risk turned out badly and contributed to the later failure of one of the blades.

The maximum shank size is determined by the thickness of the blade and the method of connection. At the time of ordering these forgings, the thickness of the blade had been determined only from aerodynamic considerations. The weight of the final blade was as yet unknown. An estimated weight was used, together with approximate aerodynamic loadings, to determine the stresses in the shank and connections. This first analysis indicated that the shanks would be strong enough as then laid out. When the blades were completely detailed, and accurate weight estimates made, and when the aerodynamic loads were more thoroughly studied, it was realized that the shank and shank-spar connection would be the weakest spot in the turbine. By this time it was too late not only to order new forgings, but even to redesign the blades. In order to use larger spars and shanks, it would have been necessary to increase the thickness of the blades.

A similar type of design decision was forced upon Jessop and Wilbur when it was found that 23⅝-inch bore main bearings were the largest obtainable. This limited them in shaft size to 23⅝ inches and, in an effort to get the utmost out

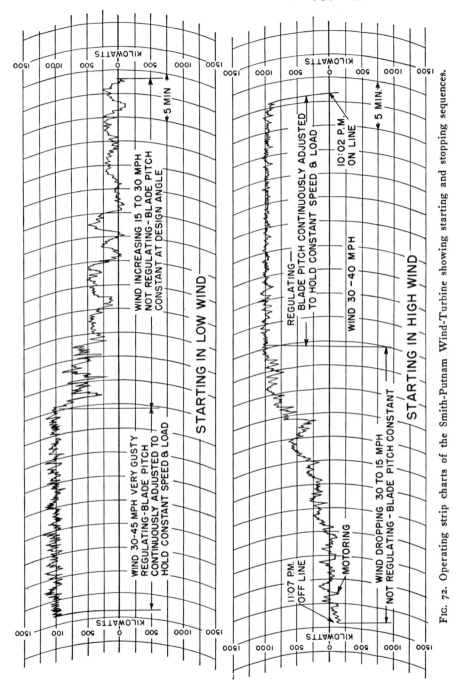

FIG. 72. Operating strip charts of the Smith-Putnam Wind-Turbine showing starting and stopping sequences.

121

of the shaft, they took the trouble to roll its surface. In the end it was found that probably there had been ample margin in the stresses of this member.

Having decided upon pitch control in response to speed changes as recorded by the Woodward governor, it became necessary to introduce an element which would permit speed change. As explained, this requirement called for a coupling between the high-speed end of the main gear and the generator; and in 1941 the only hydraulic coupling available was one which had never been continuously operated at a rating equivalent to a generator output of 1250 kilovolt-amperes. In the end, this coupling required some special cooling but otherwise gave no trouble.

The original basic design called for the hinge pins about which the blades coned to be co-axial on the center line of the main shaft. When studies were made to determine the most economical shape of the A-frame and hinge mechanism, it was found that, for ease in fabrication and simplification of the mechanism, it would be desirable to separate the hinges. It was first thought that this separation could be held to about 20 inches on either side of the center line. As finally designed, however, this distance became 32½ inches. While it was thought advisable to keep this distance small, and zero if possible, we did not realize the penalties of increasing it. Actually, the yawing and pitching moments set up by coning about these hinges introduced a certain roughness of operation, so that in the preproduction model, the design of which is described in Chapter IX, the hinge lines have again been made co-axial, at the expense of some complication in the mechanism.

Fabrication

The tower and blade spars were fabricated in the Ambridge, Pennsylvania, plant of the American Bridge Company. The tower was shipped directly to the site, and the spars to the Budd Company in Philadelphia. The blades, which consisted of a stainless-steel skin supported by ribs built up of stainless-steel structural shapes, were fabricated on these spars. The completed blades were then shipped to the Cleveland Plant of the Wellman Engineering Company, who fabricated the remainder of the structure, and shop-assembled the whole, statically balancing the blades and the rotating system before shipping to Rutland in the spring of 1941.

Erection

The erection of the wind-turbine on Grandpa's Knob presented a number of unusual hauling and lifting problems. In March, 1941, the turbine was shipped by rail from Cleveland, Ohio, to the West Rutland yards of the Vermont Marble Company, where two traveling bridge cranes, of 20-ton and 15-ton capacity, were used to unload the parts. Large trailers then transported the material to the top of the mountain, 10 trips being required to complete the task.

Some loads were heavier than the legal load limit and special permission was

FIG. 73. Assembling the blades, August, 1940.

123

obtained from the State Highway Department to move the sections over the State Highway, and two of the bridges on the route had to be reinforced with temporary cribbing. When the blades were moved, crews from the power and telephone companies had to remove and replace low wires across the highways. With the cooperation of these agencies and the State Highway Patrol, the hauling of the turbine from West Rutland to the foot of the mountain took place without mishap, but on the final 2-mile stretch from the foot of the mountain to the top, we were not so fortunate.

The grade on this 2.2-mile road, begun in August, 1940, and completed in six weeks, approached a maximum of 15 per cent and in no place was less than 12 per cent. We used a half-track in front of the truck and a bulldozer pushing or pulling, as occasion demanded. About 1000 feet below the summit, at a sharp hairpin turn on a steep grade, the pintle girder, weighing about 43 tons, including the parts already assembled on it, broke its lashings and turned over in the ditch by the road, fortunately without serious damage. Three weeks of rigging and blocking were required to get this piece back on the trailer for the remander of the trip to the top.

The tower foundations had been started in the fall of 1940, as soon as the road was completed. By the early part of December the foundations were in, and the tower was erected during the next two months, this work continuing in temperatures as low as 18 degrees below zero and in winds of 60 miles per hour. Hauling of the wind-turbine parts from West Rutland to the site was done from March 15 to May 1 of the spring of 1941, in a race against spring floods and thawing roads.

The blades were left at the bottom of the hill until all was ready for them at the summit. The pintle girder, with the main shaft assembled on it, was hoisted into place on May 15, 1941. Erection proceeded throughout the summer and the first blade was hoisted into place early in August. After erecting the first blade, it was necessary to rotate the turbine through 180 degrees, lifting the first blade to the vertically upward position, so that the second blade could be put into place (Fig. 73).

Miscellaneous blocking and rigging were then removed and on August 29, 1941, the blades were rotated by wind for the first time.

SUMMARY

The functional specification of the design was determined in 1937 and the first attempt to discover the best proportions was made in 1939. S. Morgan Smith Company accepted the basic design in 1939, and decided upon the proportions for the test unit, which was engineered under Wilbur's direction, fabricated, erected, and wind-driven for the first time on August 29, 1941, seventeen months after the decision to go ahead, and twenty-three months after the decision to explore the problem of large-scale wind-power.

Chapter VIII

TEST AND OPERATION OF THE SMITH–PUTNAM WIND–TURBINE, 1941–1945

Following the adoption of the basic design; its engineering under Wilbur's direction; and its erection on Grandpa's Knob, described in the previous chapter, a test program was carried out, from October 19, 1941, to March 3, 1945 (Ref. 27-B), culminating in the routine operation of the unit as a generating station on the lines of the Central Vermont Public Service Corporation, in the spring of 1945. Operation of the unit ended on March 26, 1945, when a blade failed.

Test Program

A test program was laid out to achieve the following aims:

1. Adjust the operation of the test unit until it should run satisfactorily, when it would be turned over to the Central Vermont Public Service Corporation as a generating station.
2. Measure the aerodynamic and mechanical efficiencies.
3. Measure the stresses, to provide a basis for a production design.

The first aim was achieved, the second approximated, and the third partially approximated (Ref. 22-B).

The risk to observers from flying ice was minimized by nestling the control house into the side of the hill and giving it a heavy concrete roof reinforced with I-beams.

By means of electrical telemetering devices (Ref. 26-B), nineteen indications of pressure, motion, temperature, and electrical output were brought to a single panel in the control house. These were:

1. Coning angle.
2. Wind direction relative to turbine shaft.
3. Up-wind coning damping pressure.
4. Down-wind coning damping pressure.
5. Servomotor pressure to increase blade pitch.
6. Servomotor pressure to decrease blade pitch.

7. Accumulator oil pressure.
8. Yaw motor oil pressure.
9. Blade pitch position.
10. Blade pitch limit.
11. Speed adjustment.
12. Up-wind gear-oil temperature.
13. Down-wind gear-oil temperature.
14. Accumulator-oil temperature.
15. Turbine speed.
16. Wind velocity.
17. Blade position.
18. Kilowatts.
19. Reactive kilovolt-amperes.

In addition this panel carried a clock, a sweep second-hand, and a date indicator (Fig. 74).

The panel containing these instruments was photographed on 16-mm. film by two cameras operating at speeds of 8 frames per minute and 8 frames per second, respectively. The slower camera was operated continuously when the turbine was in operation. The high-speed camera was run only on occasions when more detailed records were desired.

After processing, the films were viewed in a microfilm news reader or projected on a screen with a 16-mm. projector.

A Foxboro strain-gauge recorder was used to estimate the important stresses. Strain gauges were mounted on the torque tubes for the purpose of investigating flutter in the blades, and were later mounted on the pintle shaft and on a shaft in the yaw mechanism (Ref. 28-B).

From September 29 to October 19, 1941, the unit was gradually brought up to "speed-no-load." At intervals, the blades were checked for flutter by means of the strain gauges on the torque tube and the Foxboro recorder. Dr. J. P. den Hartog supervised this investigation and pronounced the blades free from flutter. After the unit had been brought up to speed, the next step was to adjust the governor so that the speed might be controlled closely. At the same time the electrical circuits were being checked and the individual parts of the switchgear tested. Two weeks were spent in drying out the generator and on October 19 the generator was run on short circuit about one and one-half hours.

Later that same day, at 6:56 P.M., the unit was phased-in to the lines of the Central Vermont Public Service Corporation for the first synchronous generation of power from the wind. The unit was run until 8:35 P.M. carrying loads which varied from 0 to 700 kilowatts. During this time the wind velocity varied from 15 to 26 miles per hour from the northeast, and was gusty (Ref. 29-B).

As soon as the turbine had been run it was found that vibration, in both the horizontal and the vertical plane, was greater than was expected. The forces causing motion in the horizontal plane, that is, about a vertical axis, are yawing forces; those causing motion in the vertical plane, that is, about a horizontal axis parallel to the plane of rotation, are pitching forces. Pitching is used here in the nautical sense and has no reference to the pitch of the blades.

The motion produced by the pitching forces was an elastic deformation of the pintle shaft and pintle girder, and sometimes amounted to 3 inches at the

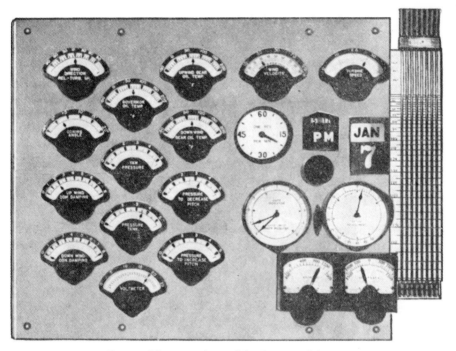

FIG. 74. The control panel in the control house.
This panel was under photographic observation
by a low-speed camera and a high-speed camera.

up-wind end. The yaw motion was quite different. The turbine was yawable about a vertical axis by means of a hydraulic motor and gearing. This transmission consisted of bevel and spur gears, together with a torque-limiting clutch. It was found that the vibrational forces in yaw, in combination with the component of torque which resulted from the inclination of the axis, produced unbalanced oscillations, the net result of which was to turn the turbine out of yaw to the left when facing up-wind. Wilbur decided, as the first of a possible series of steps to remedy the condition, to insert an irreversible worm gear in the transmission. While this eliminated the progressive creep, the oscillations in yaw were left and these yaw forces were now carried through to the tower, which would absorb the forces to the limit of the torque-limiting clutch in the yaw mechanism.

Then the tower would snap back, through as many as several inches at a radius of 6 feet.

The next step was to remove the worm gear and replace it with a three-to-one reduction, at the same time removing the torque-limiting clutch. With the addition of check and relief valves in the hydraulic system, the yaw mechanism now held the turbine in position, although with a good deal of residual oscillation in yaw. A program was instituted to discover and remove the causes of the remaining vibration. This program was twofold. First, the effect of adjustments on the test unit was investigated. Second, a theoretical analysis was made of the factors affecting the oscillating forces on the structure. The theoretical analysis is discussed in Chapter X. The adjustments were made on the coning-damping system.

The coning motion of the blades was determined from the film records taken under various conditions of operation. It was found that the amplitude of coning was less than that predicted by theory and also that the mean coning angle was less. The coning-damping mechanism was adjusted to give less damping. The amplitude of coning then increased somewhat, but the mean position did not shift. This softening of the coning damping reduced but did not eliminate the vibrational motions in yaw and in pitch. The coning links between the blades and the damping mechanism were then lengthened to allow more negative coning, and the damping was then further softened and readjusted until the vibration was reduced to a minimum, as recorded by the strain gauges. Operation then became satisfactory, although some small amount of oscillation remained (Ref. 30-B).

The determination of the aerodynamic efficiency of the turbine is complicated by the determination of the effective wind velocity. Winds with the same hourly average may give widely different values of output, depending on the actual distribution of velocity during the hour. Instantaneous readings at one anemometer are not necessarily representative of integrated values over the disc area of a wind-turbine.

A long-term statistical study is necessary to determine how the output varies with variations in wind velocity, but the fragmentary data which have been worked up show agreement with von Kármán's predictions based on theory.

A dynamic analysis of the instrumentation showed serious time lag and amplitude distortion in the readings. They had no effect on the adjustment of the test unit, but more accurate values were necessary in the investigation of blade motion, an important part of the design studies of the production unit.

The yawing and pitching forces as determined by strain gauges on the yawing mechanism and on the pintle shaft agreed with the theoretically predicted values.

Although the test program was interrupted many times by mechanical failures, as will be described in the next section, the unit was finally brought to a satisfactory operating condition after about 1100 hours of test running, spread over four and a half war years (Ref. 31-B).

Debugging the Test Unit

The troubles attending the birth of this new design were many and severe. Most of the various oil seals leaked oil out of the hydraulic systems. The packing glands on the pitching cylinder, also those on the oil tubes and on the down-wind end of the main shaft had to be entirely rebuilt. The oil head on the main shaft had to be rebored to take multiple-chevron packing. It was necessary to add oil seals to the shafts in the gear box.

The down-wind generator bearing overheated and had to be returned to Schenectady where it was redesigned by increasing the thrust clearance from 10 to 40 mills and by cutting a new oil groove.

As soon as the unit was started up for the first time, creaking noises were heard in the down-wind end of the turbine. With stethoscopes the noises were traced to the hub-post. A few loose rivets were found. The entire hub-post was then field-welded wherever accessible. It was also found that the hub-post had not been pulled up tight on the tapered shaft, because of an error in measurement during shop assembly. When the hub-post was pulled tight on the shaft, there was no more trouble with these noises.

The test unit was protected by a great many safety devices, most of which consisted of mercury-type switches, which, because of the motion of the structure, would frequently slop over, producing shut-down. With no target indication to tell which relay had caused the shut-down, it sometimes meant a long hunt to find what had happened. All of the mercury switches were replaced by mechanical switches with manually reset target indicators.

At the time when the design of the test unit was frozen, the only hydraulic coupling available for service between the gear and the generator was one whose largest model had never been operated continuously at a rating corresponding to full load in our application. It was soon found that a few hours at full load heated the coupling beyond its allowable limits.

Air-duct cooling was tried and found inadequate even in an ambient of minus 25 degrees Fahrenheit. The coupling was jacketed and sprayed with a solution of ethylene glycol in water. The freezing point of this solution was 55 degrees Fahrenheit below zero. Two radiators were hung below the pintle girder to serve as heat exchangers. The cooling solution was pumped from a sump tank to the spray nozzle, whence it drained by gravity through the radiator to the sump. In the short period during which this device was in use (March 3 to March 26, 1945) it gave satisfactory results. It was required on the test unit, although in a production unit, using an electric coupling, there would be no necessity for such special cooling.

The vertical shaft in the yaw-mechanism, which carried a pinion in its lower end meshing with the bull gear, acted as a shear pin and sheared off several times under the stress of the unexpectedly large yawing moment. A new type of yaw-mechanism was designed, in which the vertical shaft carried on its upper end a worm gear meshing with a worm free to move axially under the restraint of a damping cylinder. Spring loading of the worm kept it centered in its normal position. The final adjustments to coning damping, described elsewhere, reduced the yawing moments substantially. It was felt that this adjustment, in conjunction with a redesign of the A-frame for production, described in Chapter X, whereby the hinge pins would be co-axial and on the center line of the main shaft, would render this redesigned yaw gear unnecessary.

In May, 1942, after some 360 hours of operation, a routine inspection of the blades showed cracks in the blade skin near the roots of the blades. These cracks were concentrated over the spar, leading to the belief that the loads were being carried to the root by the skin rather than being carried into the spar by the ribs. One of the subtlest design problems is the distribution of loads through complex structures and evidently the assumptions made in the case of the blade had not been entirely valid.

If the loads were coming in in the manner indicated by the cracks, there was only one thing to do and that was to make the section strong enough to take them. To this end a heavy box was built up at the root section, of transverse and longitudinal bulkheads covered with a comparatively heavy skin. Some minor cracks developed after this repair. These further cracks were repaired by arc welding. The effect of this repair on the shank spar connection will be discussed in the section of blade failure.

The really serious interruption to the test program was not apparently a design failure but rather the failure of one of the main bearings, for unknown reasons. In February, 1943, a routine inspection discovered the down-wind main bearing running hot. It was found that the bearing had moved on the shaft and was rubbing on the end-plate. Further inspection showed the inner race cracked through. Replacement of this bearing was a serious jolt to the project. In the first place, it took two years to secure a new bearing. In the second place, installing it meant a major disassembly job. The main shaft was disconnected at the up-wind end and the turbine was allowed to drop down against temporary supports while the bearings were pulled off the exposed shaft. The new double-row spherical roller S.K.F. bearing was pulled back onto the down-wind position on the shaft, following which the up-wind main bearing and Oldham coupling were reassembled, the turbine tipped back into place and the unit placed in operation again on March 3, 1945, twenty-five months after the bearing failure.

Although the test program was interrupted by the necessity of correcting the usual number of minor design defects and although it was further interrupted by the long outage due to the failure of the main bearing, it was possible in the 1100 hours of actual operation under test to smooth out the unit and get it operating satisfactorily as a routine generating station. No evidence was found that the mechanical and aerodynamic efficiencies were not equal to the design estimates.

By 1944 enough experience had been obtained, including strain-gauge recordings, to permit a recalculation of loadings. This was carried out by Wilcox, in collaboration with Holley, under the direction of Wilbur, who concluded that the actual loadings were somewhat higher than those which had been used in the design. His anxiety increased about the blade-root sections, which had originally been very highly stressed. Accordingly, in December, 1944, and just before the unit was to go back on the line as a routine generating station, Wilbur proposed to the S. Morgan Smith Company that the test unit be torn down as soon as it had served its purpose. This decision was accepted by S. Morgan Smith Company.

Routine Operation as a Generating Station

At a planning conference held in Rutland in January, 1945, it was decided that when the turbine was again ready to run, it would be operated as a generating station, with no interruptions for routine testing. The bearing repair was completed on March 3, 1945, and the turbine was then run for the first time since February 21, 1943. Three shifts of operators were supplied by Central Vermont Public Service Corporation, and three shifts of inspectors by the S. Morgan Smith Company as a temporary precaution, until we could be sure that the turbine was functioning properly. The hydraulic coupling cooler and its controls were untried at this time. The new bearing had to be watched carefully during its run-in period, and various other new controls were in use for the first time.

The turbine was operated without incident through an unusually windless three weeks in March; during this time, it generated 61,780 kilowatt-hours in 143 hours and 25 minutes of operation, at an average level of 431 kilowatts.

The Blade Failure

On March 26 the midnight to 8 A.M. shift came on duty to find only about 5 miles per hour of wind. About 2:30 A.M. the wind freshened, and at 2:50 A.M. there was sufficient wind to start the unit. The unit was phased-in to the line at 2:55 A.M., when it was carrying from 50 to 475 kilowatts of load.

At 3:10 A.M. Harold Perry, the erection foreman, was aloft, standing on the side of the house away from the control panel and separated from it by the 24-inch rotating main shaft. A shock threw him to his knees against the wall. He started for the controls, but was again thrown to his knees. He tried again, and again was thrown down. Collecting himself, he dove over the rotating shaft, reached the controls, and, overriding the automatic controls which were already functioning, he brought the unit to a full stop in about 10 seconds by bringing the remaining blade to full feather.

One of the 8-ton blades had let go when in about the 7 o'clock * position, and had been tossed 750 feet, where it landed on its tip (Fig. 2).

* Subsequently computed by von Kármán.

It was estimated that the turbine, with one blade remaining, had made about 3 revolutions at full speed, and about 4 revolutions at diminishing speed.

Visual damage to the standing structure appeared to be confined to the leading edge of the remaining blade, which had coned negatively into one of the stops on

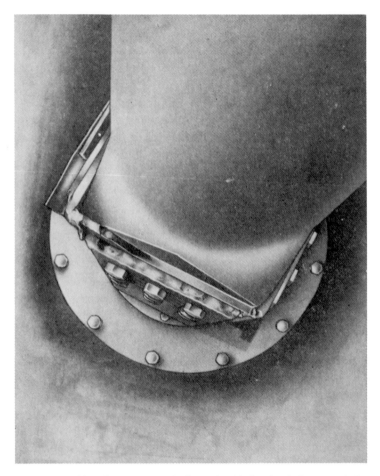

FIG. 75. Looking from outboard along the blade shank in toward the axis of rotation. The broken metallic surfaces show the members that failed.

a tower leg, but without damaging the tower. Measurements indicated that neither the main shaft nor the pintle shaft was sprung. Nevertheless, the very high unbalanced forces may have caused damage not visible.

Inspection showed that the blade spar had failed along multiple and corroded cracks just outboard of the bulkhead, itself just outboard of the bolted shank-spar connection (Fig. 75). Subsequent inspection showed similar corroded cracks well developed at the same point in the remaining blade spar.

Without being able to evaluate them, we can list the following causes which directly contributed to the failure of the spar:

A. Stress concentrations produced at the point of failure, because of an abrupt change of section in the box spar.

B. Additional stress concentrations at the point of the abrupt change in section, brought in by the bulkhead located just outboard of the change in section.

C. During the course of some field modifications in July, 1942, both the blade skin over the root area, and the bulkhead located at the point of stress concentration mentioned in A and B, were stiffened, tending to accentuate the stresses at this section.

D. At the same time, acute additional stress concentrations at this point were caused by the notch-effect of an undercut field weld, fastening the bulkhead to the four outside plates of the box spar, on the outboard face of the bulkhead. It was just at this point that failure occurred.

E. Some fatigue due to 1,500,000 revolutions of the turbine, with stress variations (but not complete reversal) in each revolution.

F. An added source of fatigue has been suggested by von Kármán, who points out that during the 2-year outage, the blades, positioned vertically, were locked in rotation and in pitch, although free to yaw. He believes that each blade was subject to continuously varying wind forces, causing it to wave like a fishpole, and setting up stress-reversals in the root sections. This motion was frequently observed during the outage and it was the upper, and more exposed blade, that failed. This stress condition is discussed in Chapters X and XII, where it is recognized as being one of the heaviest of the design loading assumptions.

SUMMARY AND CONCLUSIONS

A major structure of novel design had been engineered. In about a thousand hours of operation under test it had been concluded that the basic design was sound. In this rather short time the operation of the test unit became satisfactory as a generating station. It was operated without incident for one month by the Central Vermont Public Service Corporation as a routine generating station. At this point a structural failure occurred in a member which was known to be weak, and which had long since been redesigned for production.

The engineers of the S. Morgan Smith Company feel that they are now in a position to design a large wind-turbine with confidence, and with further improvements which will make the operation smoother, the maintenance simpler and the energy cost less.

Chapter IX

THE BEST SIZE FOR A LARGE WIND–TURBINE

It seems desirable to describe in an orderly sequence the evolution of our ideas regarding the best size for a large wind-turbine. Accordingly, in this chapter there are summarized four successive studies of the most economical dimensions, made in 1937, 1939, 1943, and 1945.

The First Approximation

In 1937 I was the first to carry out, so far as I know, a comprehensive determination of the most economical dimensions of a large wind-turbine (Ref. 10-B). This study was based on the functional specifications described in Chapter VIII.

Wilbur had made a preliminary stress analysis and Jackson and Moreland had prepared sketches, but the design was in no sense a detailed one. Rossby furnished an estimate of the vertical distribution of the wind velocity above the summit of a ridge in order to arrive at a preliminary notion of the advantages, if any, to be gained from very tall towers. The Bethlehem Steel Corporation furnished weight and cost estimates of a great variety of four-legged towers, with wide variations in the assumed loadings. Only a brief study of the increase of tower cost with increasing height was necessary to show that the increase in available energy with increase in height, as predicted by Rossby, would be insufficient to pay for very tall towers. It was clear that the best tower height over a very wide range of turbine ratings would lie somewhere between 100 feet and 300 feet.

The General Electric Company furnished weight and cost estimates of some seventy-five gears and generators covering a wide range of ratings and speeds. Similarly, the Budd Manufacturing Company and other sources supplied estimating figures of the variations in blade weight and cost with variations in radius and rating.

All costs were quoted in lots of *100 developed units*. These quotations were combined to obtain the installed costs of units rated from 250 kilowatts to 1000 kilowatts, with diameters varying from 140 feet to 320 feet. Tower heights were studied in the range from 50 feet to 300 feet. Outputs were computed from a hypothetical velocity-frequency curve, with a mean annual velocity of 25 miles an hour, thought to be typical of 3000-foot hilltops in western Massachusetts. Annual charges were set up by Jackson and Moreland at 12½ per cent. About one hundred such computations were carried out, from which were selected by inspection the designs in each sub-group that gave the lowest energy cost. These selected points were plotted in Fig. 68.

The curves of Fig. 68 show that the minimum cost is found when a generator with a capacity of 600 kilowatts is driven by a turbine with a diameter of 226 feet on a tower about 150 feet high in a wind regime with a mean annual velocity of 25 miles an hour. It was estimated that such a wind-turbine would produce energy at the switchboard at the foot of the tower at a cost of 1.52 mills per kilowatt hour. This unit would cost about $72 per kilowatt installed and would weigh about 350 pounds per kilowatt, including foundation steel.

It is interesting that the energy cost from a 1000-kilowatt unit with a diameter of 300 feet is only 7.2 per cent greater than the least cost, found at the rating of 600 kilowatts with a turbine 226 feet in diameter. The extremes of turbine diameter and generator rating which may be had within a variation from the minimum energy cost of 10 per cent are found to be 160 to 300 feet and 300 to 1200 kilowatts. This insensitivity of the cost envelope over a rather wide range of diameters and ratings was found to be characteristic of the three later approximations.

Another characteristic of this curve also found in all later ones is that the energy cost increases more quickly when the diameter is reduced below the best value than it does when the diameter is increased above it. (Compare Figs. 69, 76A, and 77.)

The cost of the best 1000-kilowatt unit and the best 500-kilowatt unit is less than that of the best 750-kilowatt unit, indicating some discontinuity in the quotations, probably in the gear or generator price series. Discontinuity in gear cost alone is expected when the number of steps in the gear is changed. Mass production of certain sizes of equipment lowers the cost in steps and introduces discontinuities in the otherwise smooth curves of cost plotted against rating.

This first approximation named 600 kilowatts as the most economical capacity, while later studies showed that 1500 kilowatts was the most economical. At the time this study was made, in 1937, the gear and generator were thought to comprise a larger percentage of the total cost than has since been found to be the case. This error arose because of my lack of detailed knowledge of the rest of the structure and of my belief that it would be lighter and more simple than it proved to be when finally built.

As the cost of the gear and generator decreases relative to the rest of the structure, it is obvious that the most economical generator size will increase, while the most economical diameter will decrease. These relations have been confirmed in later approximations.

The Second Approximation

In Chapter I, it was explained that the first step taken by the S. Morgan Smith Company, after having approved my basic design, was to review my estimate of

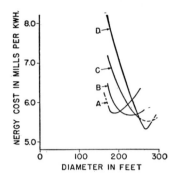

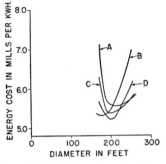

FIG. 76A. Third approximation of the most economical dimensions. Variation in the energy cost with variation in turbine diameter for various generator ratings, but at a constant rotational speed of 1.5 radians per second.

Curve A   1000 kilowatts
Curve B   1500 kilowatts
Curve C   2000 kilowatts
Curve D   2500 kilowatts

FIG. 76B. Third approximation of the most economical dimensions. Variation in the energy cost with variation in turbine diameter for various generator ratings, but at a constant rotational speed of 2.0 radians per second.

Curve A   1000 kilowatts
Curve B   1500 kilowatts
Curve C   2000 kilowatts
Curve D   2500 kilowatts

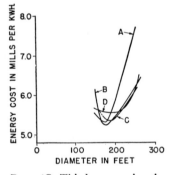

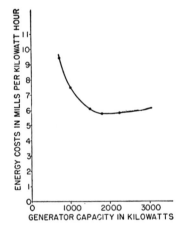

FIG. 76C. Third approximation of the most economical dimensions. Variation in the energy cost with variation in turbine diameter for various generator ratings, but at a constant rotational speed of 2.5 radians per second.

Curve A   1000 kilowatts
Curve B   1500 kilowatts
Curve C   2000 kilowatts
Curve D   2500 kilowatts

FIG. 77. Fourth approximation of the most economical dimensions. Variation in the energy cost with variation in the capacity of the wind-turbine generator, in a block of 9000 kilowatts of capacity installed on the 4000-foot Lincoln Ridge in Vermont. The plotted values come from line 13 of Table XIII.

136

the cost of producing energy from the wind by means of this design (Ref. 24-B).

The general method used in the first approximation was adopted for the second, which was carried out in 1939 by a special computing staff created for the S. Morgan Smith Company at the Budd Manufacturing Company under the general direction of Benjamin Labaree. The wind regime was specified by Petterssen who estimated a mean annual velocity of 28.8 miles an hour as being typical of good wind-turbine sites in Vermont. The outputs for a large number of turbines were computed by the special staff, based on a report by von Kármán (Ref. 32-B), who specified the dimensions and characteristics of a series of blades (Fig. 78) designed to give maximum efficiency in a wind speed of 14 miles an hour at a tip-speed of 165 feet per second. Von Kármán had computed the variation in the power output of these blades with varying wind speeds and these computations had been verified by Elliott Reid, who had tested models in the wind-tunnel at Stanford University for the S. Morgan Smith Company (Ref. 33-B). Finally, the Budd Company devoted several thousand man-hours to the preparation of accurate cost estimates of a representative series of these blades.

Again, master computation charts were laid out, but we were able to profit by the work which had been done since 1937, permitting better weight estimates. We investigated diameters between 150 and 300 feet and generator ratings between 750 and 3000 kilowatts, at speeds of 600 and 900 revolutions per minute, and some of the weights were re-estimated by Wilbur.

Tower height and turbine speed were each held constant at values derived from the first approximation. We now know that these two values were each a little low.

A total of 122 computations was carried out. In addition, output only, without cost, was determined for 294 other combinations of diameter, rating, and speed (Ref. 34-B). Jackson and Moreland again set the annual charges at 12½ per cent. The total costs were summed, the energy cost computed in mills per kilowatt-hour, and the low cost design of each sub-group was again plotted, yielding the curves of Fig. 69. It is seen that the most economical unit has a diameter lying between 175 feet and 235 feet, based on blade type 2-M (Fig. 78), and a generator capacity lying between 1000 kilowatts and 2000 kilowatts, at a speed of 600 revolutions per minute.

Within these limits the energy cost at the switchboard at the foot of the tower does not vary by more than 2 per cent. It seemed likely that the design to be chosen for ultimate production should be selected from the upper or right-hand end of the range, which would mean a turbine with a diameter of at least 225 feet and a capacity of at least 2000 kilowatts. The reason for this is that the second approximation was based on the energy costs of a single unit and did not reflect the effect of such items as access road, connecting transmission line, and erection costs. The effect of charges of this type is to make it desirable to occupy the finite number of sites on a ridge with large rather than small units.

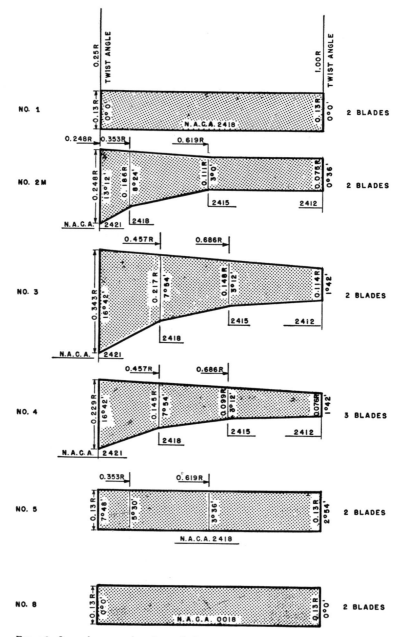

FIG. 78. Second approximation of the most economical dimensions. The series of eight blades studied.

On the other hand, the dimensions of the experimental unit were selected from the lower or left-hand end of the flat envelope, since the S. Morgan Smith Company wished to experiment with the smallest unit which would, nevertheless, be dynamically representative of the ultimate production unit. Accordingly, as explained in Chapter VII, it was decided, as a result of the second approximation, that the test unit should be rated at 1250 kilowatts, with a diameter of 175 feet, a generator speed of 600 revolutions per minute, and a hub height of 125 feet.

The installed cost of the most economical unit, as determined in the second approximation, was about $56 per kilowatt, and the weight, including foundation steel, was about 150 pounds per kilowatt, *based on lots of 100 developed units*.

The Third Approximation

In 1943 a large wind-turbine had been tested in operation for two years. The staff of the Wind-Turbine Division of the S. Morgan Smith Company had accumulated a body of experience and had been able to consider many ideas looking toward the simplification of the test unit.

Accordingly, when the War Production Board in 1943 requested cost estimates for a Victory Model to be used in certain areas where there was a deficiency of power, it was decided to take advantage of the experience gained and to redetermine the most economical dimensions (Ref. 35-B).

Because of the war, it was not feasible to obtain from vendors new smooth curves of cost versus rating for the various components of the wind-turbine. A

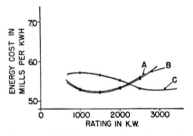

Fig. 79. Third approximation of the most economical dimensions. Variation in the energy cost with turbine rating for various turbine speeds.

Curve A   2.5 Radians per second
Curve B   2.0 Radians per second
Curve C   1.5 Radians per second

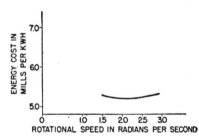

Fig. 80. Third approximation of the most economical dimensions. Variation in the energy cost with turbine speed.

substitute method was used. It was assumed that the cost of each component of the turbine varied in some regular manner with variations in turbine diameter, turbine speed, and generator capacity. For example, the cost of the blades was assumed to vary with the square of the diameter; to be unaffected by the turbine speed; and to vary with the one-half power of the generator capacity. It was recognized that actually the cost of the blades may vary as the 1.8 power or the 2.2 power of the diameter. However, it was felt that errors in such estimates should tend to cancel out since the number of items is rather large.

The capital charges were again based on the advice of Jackson and Moreland and were set at 12½ per cent. The hypothetical wind regime was based on Glastenbury and had a mean annual velocity of 21.5 miles an hour.

In Figs. 76 A, B, C are plotted the energy costs against diameter for various generator sizes and turbine speeds. In Fig. 79 selected minimum values of cost are plotted against generator capacity, and in Fig. 80 against turbine speed.

The study again indicates that the most economical diameter is about 200 feet and the capacity about 1500 kilowatts.

It is seen that these curves show all the characteristics of the curves of the first and second approximations and substantiate the findings of the second approximation rather closely.

The energy cost at the switchboard at the foot of the tower of the most economical unit was found to be 5.2 miles per kilowatt-hour (Ref. 36-B). This unit was estimated to cost, installed, $149 per kilowatt, and to weigh 450 pounds per kilowatt, including foundation steel.

The Fourth Approximation

In part because of the interest of the War Production Board and in part to make the best of the enforced delay while waiting for the new down-wind main bearing, it was decided in 1943 to move ahead with at least the preliminary stages of the formal design of a preproduction model.

No rigorous attempt had yet been made to determine in detail the most economical blade shape. The first step in the fourth approximation consisted, therefore, of a study of blade shape (Ref. 37-B).

Designs were laid out for sixteen blades, for which outputs and costs were determined for each of two diameters, in each of two wind regimes for each of several generator ratings and each of several turbine speeds.

The ideal blade is a blade in which each elemental section is operating at the maximum theoretical efficiency. This requires that the width of the blade and the blade angle shall vary continuously along the radius, and with a distribution which itself will vary with the wind-speed. Obviously, such a blade is impractical.

The ideal turbine has no profile drag, which is ordinarily ignored in computing the output by classical theory. For our purposes it was necessary to include the profile drag. Accordingly, following a suggestion by von Kármán, an extension of the classical theory was carried out by Wilcox and Holley. By means of this extension they were able to compute the output of an ideal turbine for various values of the lift-drag ratio. This ratio is a measure of the profile drag of a blade section.

The next step in determining the best blade shape was to approximate the ideal blade shape for various values of the tip-speed ratio. This is the ratio between the linear speed of the tip of the blade, in the plane of rotation, and the speed of the wind at right angles to the plane of rotation.

For any value of the tip-speed ratio there is a corresponding value of the efficiency of an ideal turbine.

Four blades were designed, the respective maximum efficiencies occurring at four different tip-speed ratios, viz., 4.0, 6.0, 8.0, and 10.0. For each of the four values of the tip-speed ratio, Wilcox and Holley used three values of the lift coefficient, viz., 0.60, 0.80, and 1.20, each held constant from root to tip.

In addition, three blades were designed having a constant blade width along the radius (rectangular plan-form). At the suggestion of von Kármán, an additional blade was designed, having a linear variation of lift coefficient from 0.6 at the tip to 1.20 at the root, for a tip-speed ratio of 6.0. These sixteen different blades are shown in Fig. 81.

The choice of the blade cross-section is independent of the other factors determining the blade shape. From the aerodynamic standpoint a high maximum value of the lift coefficient and a low value of the drag coefficient are desirable. Structurally the airfoil must be. sufficiently deep to accommodate the supporting members. Flat surfaces or convex surfaces are cheaper to build than concave surfaces. For the test unit we had chosen the N.A.C.A. 4418 section as having good aerodynamic characteristics, sufficient thickness for the internal structure, and no concave surfaces, and it was selected for this study as well.

An aerodynamic analysis was made of each of the blades (Ref. 38-B). In some of our earlier studies, the simple blade-element method of aerodynamic analysis was employed, as a useful means of approximation. In this study, however, we decided to employ the vortex theory (Ref. 11-A) of the propeller, modified for the windmill condition, as being the most accurate method available.

From the aerodynamic output so computed were subtracted the losses caused by hub-windage and tower shadow, to give the gross input to the main shaft. From this input were subtracted the gear losses, the coupling losses, the generator losses, and the power required to operate the servo-mechanisms and the auxiliary equipment, the net balance being the generator output, whose variation with wind velocity is plotted, for several of the designs, in Fig. 82.

The variation in annual output of each of the sixteen blades, with variations in the rotational speed, was determined for each of two diameters (175 feet and 200 feet), and for several generator ratings (Table XII). The output of each of these combinations was determined for two different wind regimes, one characteristic of the West Indies, with a mean velocity of 18.4 miles per hour, and one characteristic of Glastenbury Mountain (3700 feet) in Vermont, with a mean velocity of 21.5 miles per hour. Figs. 83 and 84 show 2 such sets of output curves for the 16 blades investigated. Each figure is for one combination of diameter,

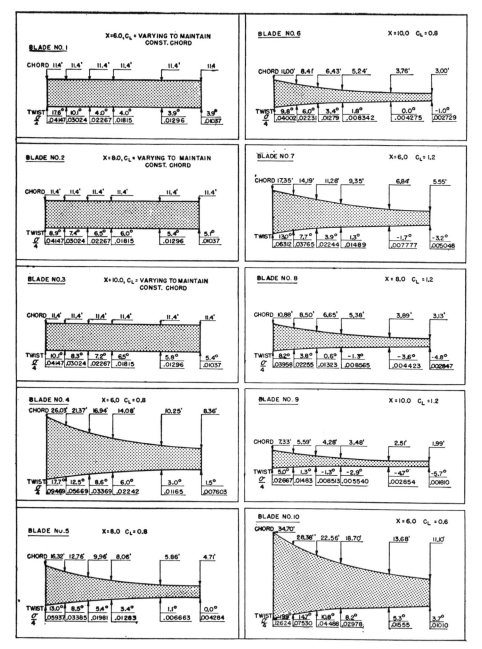

FIG. 81 (Left). Fourth approximation of the most economical dimensions. The sixteen blades which were studied.

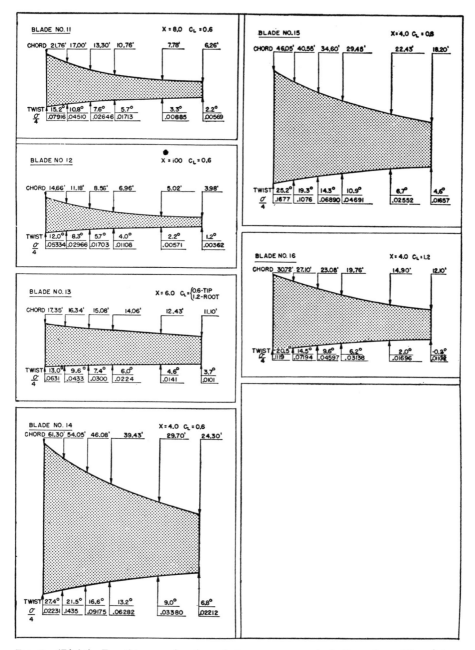

FIG. 81 (Right). Fourth approximation of the most economical dimensions. The sixteen blades which were studied.

143

wind regime, and rated power. A critical comparison of the 12 sets of curves led to the following conclusions:

1. A single type of blade can be used over a wide range of wind regimes, rated powers, and diameters. The limits of the various ranges were not determined.

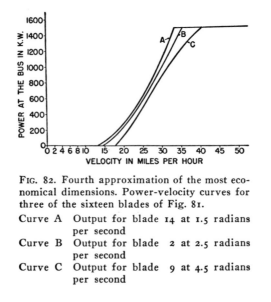

FIG. 82. Fourth approximation of the most economical dimensions. Power-velocity curves for three of the sixteen blades of Fig. 81.

Curve A   Output for blade 14 at 1.5 radians per second

Curve B   Output for blade 2 at 2.5 radians per second

Curve C   Output for blade 9 at 4.5 radians per second

TABLE XII. GENERATOR RATINGS USED FOR THE DETERMINATION OF OUTPUT

Diameter	Rating
175	1000
	1250
	1500
200	1200
	1500
	1800

2. The best blade (No. 15) gave an output which was 95 per cent of the ideal output. We made no investigation of other plan forms and twists which might have shown an improvement over this figure.

3. The largest tapered blade that it seemed reasonable to build (No. 13) gave an output which was 92 per cent of the ideal.

4. The best rectangular blade (No. 1) gave an output which was 88 per cent of the ideal, or 96 per cent of blade No. 13.

5. The gain to be achieved by using a different airfoil section is small since we are limited in our choice by structural considerations.

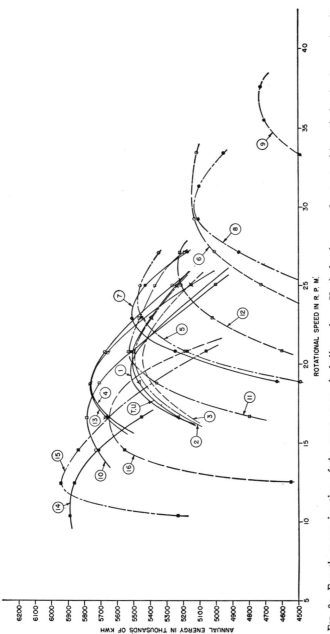
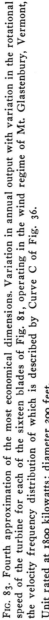

FIG. 83. Fourth approximation of the most economical dimensions. Variation in annual output with variation in the rotational speed of the turbine for each of the sixteen blades of Fig. 81, operating in the wind regime of Mt. Glastenbury, Vermont, the velocity frequency distribution of which is described by Curve C of Fig. 36. Unit rated at 1800 kilowatts; diameter 200 feet.

145

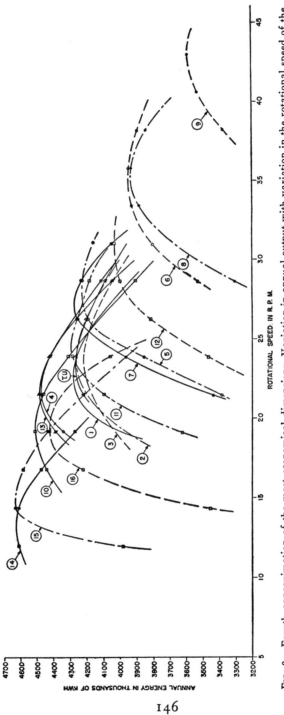

FIG. 84. Fourth approximation of the most economical dimensions. Variation in annual output with variation in the rotational speed of the turbine for each of the sixteen blades of Fig. 81, operating in the wind regime of Mt. Glastenbury, Vermont, the velocity frequency distribution of which is described by Curve C of Fig. 36. Unit rated at 1500 kilowatts; diameter 175 feet.

THE BEST SIZE FOR A LARGE WIND-TURBINE

Production costs were estimated by the Budd Manufacturing Company for the rectangular blade No. 1 and the tapered blade No. 13. The Budd Company costs are given in Fig. 85. It can be seen that for ten units (twenty blades) the tapered blade is twice as expensive as the rectangular blade.

Because the ratio between the cost of the most efficient tapered blade of practical dimensions (No. 13) and the rectangular blade (No. 1) is over 2 to 1, while the ratio in outputs is not over 1.10 to 1, *it seems unlikely that there is any combination of economically usable wind regime, turbine diameter, and generator rating for which blade No. 1 would not also be the most economical blade for a large wind-turbine.*

Now, the output from a rectangular blade is relatively insensitive to the amount of twist, that is, to the change in blade angle from root to tip. Thus the output from blade No. 1, if untwisted, is 98 per cent of the output when twisted.

If the blades are made up in stainless steel, they would be twisted, since in this type of construction twist costs nothing. But, the cost of twisting a mild steel blade is not paid for by the increase in annual output.

The economics of twist are further discussed in Chapter X.

The maximum efficiency of the most economical blade will, therefore, be 88 per cent of the theoretical maximum, if the blade is made up in stainless steel, and 86 per cent if made up in mild steel.

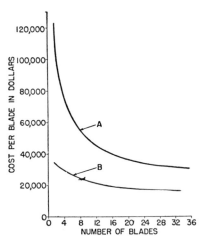

FIG. 85. Fourth approximation of the most economical dimensions. Variation in 1944 blade costs with quantity, from estimates supplied by the E. G. Budd Company.

Curve A Represents tapered blade No. 13 of Fig. 81

Curve B Represents rectangular blade No. 1 of Fig. 81

The two previous approximations of 1935 and 1939 had indicated that there would be a small but real advantage in increasing the test unit diameter by 25 feet or, possibly, 50 feet. This economic advantage was, however, not a sufficient inducement to depart in the preproduction design from the immediate range of our rather limited experience. Therefore, and also in order to simplify the calculation of what we then expected would be a preliminary cost study, it was arbitrarily decided to hold the diameter to 175 feet in the fourth approximation.

It was felt that if this economic study should prove encouraging, it would then be proper to consider the benefits to be obtained by redesigning the preproduction unit to conform more closely to the most economical dimensions.

A second primary variable, which it was arbitrarily decided to hold constant in the fourth approximation, was tower height. The vertical distribution of velocity that we had measured at Grandpa's Knob, Pond, Biddie, Seward, and Mt. Washington, left little doubt that, if a hill has the aerodynamic characteristics enabling it to speed up the wind-flow by 20 per cent or more, *the maximum velocity will not be found higher than 250 feet above the summit, and the most economical tower height would accordingly lie below 200 feet.* So we decided that the eco-

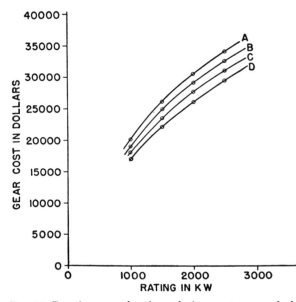

FIG. 86. Fourth approximation of the most economical dimensions. Variation in gear cost with rating, for various rotational speeds.

Curve A	25.0 revolutions per minute
Curve B	27.5 revolutions per minute
Curve C	30.0 revolutions per minute
Curve D	33.0 revolutions per minute

nomic gain to be derived from a tower higher than 150 feet, would not, in itself, be sufficient to dominate our conclusions about the economics of large-scale wind-power. Accordingly, for the fourth approximation, we held the tower height constant at 150 feet.

The hypothetical wind regime was based on our accumulated experience in Vermont and represented our expectations on Lincoln Ridge at an elevation of 4000 feet. The assumed mean annual velocity was 30 miles an hour.

With the most economical blade identified, the loadings recomputed, and simplifying assumptions made concerning turbine diameter, tower height, and mean wind velocity, the fourth and most rigorous approximation of the economically ideal size for a wind turbine was carried out in 1945 (Ref. 18-B). This approximation was based not on a site occupied by a single unit, but on a site occupied by a block of units, and it included such charges as the connecting high-line and access roads.

For the fourth approximation entirely new costs were obtained for all parts of the turbine, as will be described in Chapter XI. The items principally affecting the most economical speed and rating are the gear costs, the coupling costs and the costs of the generator and its associated equipment. Gear costs vary with input torque but are not affected by changes in the step-up ratio. Fig. 86 shows variations in the gear cost with rating.

The cost of the coupling and generator is combined and plotted against genera-

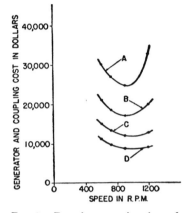

FIG. 87. Fourth approximation of the most economical dimensions. Variation in the cost of the generator and the coupling with variation in speed.

Curve A   2500 kilowatts
Curve B   2000 kilowatts
Curve C   1500 kilowatts
Curve D   1000 kilowatts

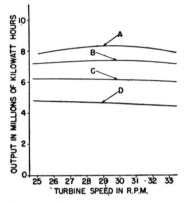

FIG. 88. Fourth approximation of the most economical dimensions. Variation of turbine output with turbine speed for various turbine ratings.

Curve A   2500 kilowatts
Curve B   2000 kilowatts
Curve C   1500 kilowatts
Curve D   1000 kilowatts

tor speed in Fig. 87. It is seen that for all generator capacities 900 revolutions per minute is the most economical speed. Since no other items affect the choice of generator speed, we selected 900 revolutions per minute and combined the generator and coupling costs at this speed with the gear costs to determine the most economical turbine speed. This is substantially correct and sufficiently close for our purposes. It is true that the weight and cost of the blades will vary somewhat with turbine speed, as will also the main shaft and certain other parts. However, these variations are minor and we do not think that the uncertainty introduced by them into the determination of the most economical turbine speed exceeds ± 0.5 revolutions per minute.

The variation of output with turbine speed for various generator ratings is shown in Fig. 88. The variations in the cost of the generator, coupling, gear and electrical equipment as the rating and speed are varied, are given in Fig. 89. The energy cost of these components is computed and plotted in Fig. 90. The best turbine speed for various ratings is found by noting the position of the minima of the

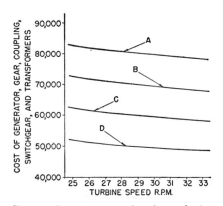

FIG. 89. Fourth approximation of the most economical dimensions. Variation in the cost of the generator, coupling, gear, switchgear and transformers, with turbine speed, for various ratings.

Curve A 2500 kilowatts
Curve B 2000 kilowatts
Curve C 1500 kilowatts
Curve D 1000 kilowatts

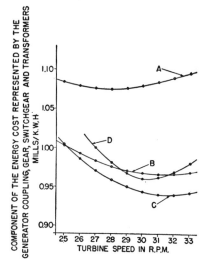

FIG. 90. Fourth approximation of the most economical dimensions. Variation in the component of the energy cost representing the cost of the generator, coupling, gear, switchgear, and transformers, with variation in turbine speed, for various ratings.

Curve A 1000 kilowatts
Curve B 1500 kilowatts
Curve C 2500 kilowatts
Curve D 2000 kilowatts

cost curves. The variation of this best speed as the generator rating varies is plotted in Fig. 91.

The value of 9000 kilowatts as the block of capacity to be analyzed was chosen because the Central Vermont Public Service Corporation was not interested in anything over 10,000 kilowatts of capacity and the number 9000 is readily divisible by various standard generator ratings.

Six combinations of generator rating and number of units were considered. The respective costs are tabulated in Table XIII and the energy costs are plotted in Fig. 77. We found the most economical capacity to lie between 1600 and 1800 kilowatts.* For the preproduction unit we selected 1500 kilowatts because this

* The outputs used in these computations were not those of Fig. 88, but corrected values thought to be more accurate.

was a standard size. By entering this value in Fig. 91, we find that the corresponding proper turbine speed is 31.5 revolutions per minute.

Thus, the fourth approximation tells us that the preproduction unit will be a turbine with two blades rectangular in plan-form; 175 feet in diameter; operating at 31.5 revolutions per minute; and driving a 1500-kilowatt generator at 900 revolutions per minute.

It will be noted that the general shape of the envelope curve of Fig. 77 is similar to that of the envelopes found in the three previous approximations and, since the dimensions of the most economical unit were found to be so nearly similar, particularly in the second, third, and fourth approximations, the *conclusion is inescapable that, for this general type of large wind-turbine and over a fairly wide range of wind regimes, the most economical dimensions are definitely known, within limits of ± 50 per cent.*

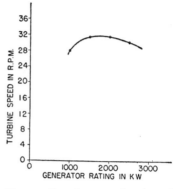

FIG. 91. Fourth approximation of the most economical dimensions. Variation between the best turbine speed and the turbine rating.

Drastic modifications in the design would, presumably, modify the values of some or all of the dimensions of the most economical unit. Thus, if the gear could be eliminated or if the blade cost could be cut in half, it would, doubtless, be found that the most economical ratings and dimensions had increased and this would presumably be the result if some of the modifications and drastic changes, discussed in the next chapter, should prove successful.

In Table XIV the four approximations are summarized and compared with a set of values illustrating the speculations of Chapter XII on how to reduce the cost of wind power.

The line of data in Table XIV describing "Fourth Approximation 1945" lists the dimensions whose determination is illustrated in Table XIII and Figure 77; the annual charges of 12 per cent are those determined as applicable to Central Vermont Public Service Corporation (p. 190 and Table XVIII); the unit cost of $191 per kilowatt and the unit weight of 497 pounds per kilowatt are taken from the detailed cost study of the preproduction unit, described in Chapter XI and summarized in Table XV; and the net salable unit output of 3320 kilowatt-hours per kilowatt per year is the value determined for the proposed installation on Lincoln Ridge (page 191).

TABLE XIII. FOURTH APPROXIMATION OF THE BEST SIZE FOR A LARGE WIND-TURBINE

	12 × 750			9 × 1,000			6 × 1,500		
	Dollars per Kilowatt	Dollars per Unit	Dollars per Installation	Dollars per Kilowatt	Dollars per Unit	Dollars per Installation	Dollars per Kilowatt	Dollars per Unit	Dollars per Installation
1. Cost of All Parts Which Vary with Capacity Includes: Gear, Generator, Coupling, Transformers, and Incremental Costs of Blades, Shafts, Bearings, Pintle Girder and Tower	$ 57.33	$ 43,000	$ 516,000	$ 45.00	$ 45,000	$ 405,000	$ 34.67	$ 52,000	$ 312,000
2. Rest of the Structure	240.00	180,000	2,160,000	180.00	180,000	1,620,000	120.00	180,000	1,080,000
Total F.O.B. Cost	297.33	223,000	2,676,000	225.00	225,000	2,025,000	154.67	232,000	1,392,000
3. Site Selection	1.11	832	10,000	1.11	1,111	10,000	1.11	1,667	10,000
4. Land	0.05	38	450	0.05	50	450	0.05	75	450
5. High-Line	14.65	10,988	131,850	14.65	14,650	131,850	14.65	21,975	313,850
6. Road	3.00	2,250	27,000	2.67	2,670	24,030	2.33	3,495	20,970
7. Erection	26.00	19,500	230,400	24.00	24,000	216,000	22.00	33,000	198,000
8. Total Installed Cost	342.14			267.48			194.81		
9. Annual Charge (10%)	34.21			26.75			19.48		
10. Annual Output (Lincoln Ridge)	4,500 Kwh.	3,375,000 Kwh.	40,500,000 Kwh.	4,108 Kwh.	4,108,000 Kwh.	36,972,000 Kwh.	3,500 Kwh.	5,250,000 Kwh.	31,500,000 Kwh.
	Mils per Kwh.			Mils per Kwh.			Mils per Kwh.		
11. Energy Cost	7.603			6.511			5.566		
12. Output Correction Factor	1.200			1.145			1.090		
13. Net Energy Cost	9.124			7.455			6.067		

TABLE XIII. FOURTH APPROXIMATION OF THE BEST SIZE FOR A LARGE WIND-TURBINE (*continued*)

	5 × 1,800			4 × 2,250			3 × 3,000		
	Dollars per Kilowatt	Dollars per Unit	Dollars per Installation	Dollars per Kilowatt	Dollars per Unit	Dollars per Installation	Dollars per Kilowatt	Dollars per Unit	Dollars per Installation
1. Cost of All Parts Which Vary with Capacity. Includes: Gear, Generator, Coupling, Transformers, and Incremental Costs of Blades, Shafts, Bearings, Pintle Girder and Tower	$ 32.58	$ 58,640	$ 293,220	$ 30.67	$ 69,000	$276,000	$ 30.38	$ 91,153	$273,459
2. Rest of the Structure	100.00	180,000	900,000	80.00	180,000	720,000	60.00	180,000	540,000
Total F.O.B. Cost	132.58	238,640	1,193,220	110.67	249,000	996,000	90.38	271,153	813,459
3. Site Selection	1.11	2,000	10,000	1.11	2,500	10,000	1.11	3,333	10,000
4. Land	0.05	90	450	0.05	113	450	0.05	150	450
5. High-Line	14.65	26,370	131,850	14.65	32,963	131,850	14.65	43,950	131,850
6. Road	2.22	3,996	19,980	2.11	4,748	18,990	2.00	6,000	18,000
7. Erection	20.00	36,000	180,000	17.78	40,000	160,000	14.34	43,000	129,000
8. Total Installed Cost	170.61			146.37			122.53		
9. Annual Charge (10%)	17.06			14.64			12.25		
10. Annual Output (Lincoln Ridge)	3,200 Kwh.	5,760,000 Kwh.	28,800,000 Kwh.	2,650 Kwh.	5,962,500 Kwh.	23,850,000 Kwh.	2,100 Kwh.	6,300,000 Kwh.	18,900,000 Kwh.
	Mils per Kwh.			Mils per Kwh.			Mils per Kwh.		
11. Energy Cost	5.335			5.523			5.835		
12. Output Correction Factor	1.072			1.053			1.035		
13. Net Energy Cost	5.712			5.816			6.039		

153

TABLE XIV. SUMMARY

The Best Size for a Large Wind-Turbine

	Mean Velocity, mph.	Annual Charges, Per Cent	Diameter, Feet	Tower Height, Feet	Generator Rating, Kilowatts	Unit* Cost, $/Kw.	Pound Cost, $/Lb.	Unit* Weight, Lbs.	Net Saleable Unit Output, Kwh./Kw.	Energy* Cost at the Bus, Mils per Kwh.
First Approximation, 1937	25	12½	226	125	600	$72	0.21	350	5900	1.5
Second Approximation, 1940	28.8	12½	200	125	1500	56	0.37	150	4400	1.6
Third Approximation, 1943	21.5	12½	200	(150)**	1500	149	0.33	450	3575	5.2
Fourth Approximation, 1945	26	12	(175)**	(150)**	1600	191	0.39	497	3320	6.9
Based on Chapter XII										
(a) Private Capital	30	12½	235	165	3000	100	0.33	300	4100	3.0
(b) Government Capital	30	8	235	165	3000	100	0.33	300	4100	2.0

* Installed, but exclusive of transformers and connecting transmission line.
** Arbitrarily held to this value.

TABLE XV. COST SUMMARY OF PREPRODUCTION UNIT, 1945

	Dollars	Dollars per Kw.	Weight	Pounds per Kw.	Dollars per Lb.
1. Engineering	$10,500.00	7.00			
2. Manufacturing					
2.1 Standard Items Now in Production					
2.1–1 Generator	8,870.00	5.91	20,250.00	13.50	0.438
2.1–2 Main Gears	20,344.00	13.56	29,000.00	19.33	0.702
2.1–3 Coupling Electric	4,612.50	3.08	9,700.00	6.47	0.476
2.1–4 Governor	2,508.00	1.67	2,000.00	1.33	1.254
2.1–5 Bearings	16,282.00	10.85	11,500.00	7.67	1.416
2.1–6 Switch Gear	5,125.00	3.42	4,000.00	2.67	1.281
2.1–7 Couplings (Flexible)	1,970.00	1.31	5,500.00	3.67	0.358
2.1–8 Elevator	2,665.00	1.78	8,000.00	5.33	0.333
2.1–9 Service Hoist	1,680.00	1.12	5,000.00	3.33	0.336
2.1–10 Miscellaneous Electrical	2,100.00	1.40	6,500.00	4.33	0.323
2.1–11 Tower (Includes Erection)	21,395.00	14.62	180,000.00	120.00	0.122
2.1–12 Paint	691.00	.46	1,500.00	1.00	0.461
2.2 Items Peculiar to the Smith-Putnam Wind-Turbine					
2.2–1 Blades	29,480.00	19.65	67,000.00	44.67	0.440
2.2–2 Hub Assembly	42,935.00	28.67	155,000.00	103.33	0.277
2.2–3 Pintle Assembly	48,600.00	32.40	200,000.00	133.33	0.243
2.2–4 Patterns, Tools, Jigs, and Fixtures	800.00	.53			
Total F.O.B. Cost	221,097.50	147.40	704,950.00	469.96	0.314
Contingency 10%	22,109.75	14.74			
TOTAL	243,207.25	162.14	704,950.00	469.96	0.345
3. Installation					
3.1 Freight**	2,054.00	1.37	437,000.00		0.0047
3.2 Land	0.00	0.00			
3.3 Road	7,460.00	4.97			
3.4 Erection (Includes Foundation Steel)	30,000.00	20.00	40,000.00*	26.67	0.040*
Total Installed Cost	260,611.50	173.74	744,950.00	496.63	0.350
Contingency 10%	26,061.15	17.37			
TOTAL	286,672.65	191.11	744,950.00	496.63	0.385
4. Connection					
4.1 Transformers	3,600.00	2.40	15,000.00	10.00	0.240
4.2 High Line	15,000.00	10.00			
Total Cost of Unit (Connected to Existing System)	279,211.50	186.14	759,950.00	506.63	0.367
Contingency 10%	27,921.15	18.61			
TOTAL	307,132.65	204.75	759,950.00	506.63	0.404

* Foundation steel only.
** Only those items shipped from York, Pennsylvania.

SUMMARY

1. Three determinations of the most economical size for a wind-turbine of my 1937 design were carried out in 1940, 1943, and 1945, respectively. The sum of the evidence indicates that the best dimensions of this design will fall within the following limits, over a fairly wide range of wind regimes, and assuming a mountain-top location:

Tower height	150–175 feet
Generator rating	1500–2500 kilowatts
Turbine diameter	175–225 feet

2. The unit weight of the fourth approximation was 497 pounds per kilowatt. (The test unit weighed 500 pounds per kilowatt.)

3. The mean pound price had nearly doubled in eight years, rising from about $0.21 per pound in 1937 to $0.39 per pound in 1945.

Chapter X

THE DESIGN OF A PREPRODUCTION MODEL, 1943–1945

Introduction

When it was found that the third approximation had verified the best size of a large wind-turbine, as found in the two previous approximations, the design of a preproduction model was begun.

The design of any structure, for which the forces associated with the acceleration of gravity are significant, must necessarily proceed by a series of approximations. In the first approximation it is usual to begin by arbitrarily assuming some distribution of the mass. Next, the forces associated with this mass distribution are calculated; then the non-mass forces. Finally, a stress analysis is carried out. A design based on this stress analysis yields a new mass distribution. For this new mass distribution, the above cycle is repeated. Successive cycles are carried through until all members of the structure are designed with consistent strength.

In such structures as bridges or buildings, for which a great deal of previous experience is available, the first estimate of mass distribution usually agrees closely with the final distribution. For these structures, one, or at the most two, cycles will suffice. The wind-turbine differs from these structures in that, first, there is no backlog of experience, and second, the mass forces in the wind-turbine depend largely on accelerations which in turn depend on the mass distribution. For this reason, one must carry through a new dynamic analysis for each new mass distribution, and several successive approximations will be required.

When the test unit was designed, there was not time, as described in Chapter I, to carry out a sufficient number of such cycles to produce a design which would incorporate consistent strength with maximum economy. Furthermore, the meager knowledge in 1940 of the probable wind conditions to be encountered did not justify a precise analysis and design.

When, in 1944, it was decided to start the design of a preproduction unit, our knowledge had been substantially augmented by actual experience with the test unit and by certain theoretical studies carried out during the test period. It was the intention to carry through as many design cycles as necessary. But the project was abandoned in the early stages of this program, the first cycle of which was not completely carried out.

Characteristics of the Preproduction Unit

The third economic study to determine the best size of a large wind-turbine had fixed some of the characteristics of the preproduction unit, as described in Chapter IX. These were as follows:

1. Diameter—175 feet
2. Turbine Speed—2.71 radians per second, or 26 revolutions per minute
3. Generator Capacity—1500 kilowatts
4. Coupling Type—electric, torque-limiting
5. Blade Plan-form—rectangular
6. Blade Chord—11.4 feet
7. Blade Length—65.6 feet
8. Blade Section—N.A.C.A. 4418
9. Blade Twist—5°, discontinuously, in 4 steps.

During the test program an analysis was carried out to determine the factors affecting the oscillating forces on the structure (Ref. 39-B) described in Chapter VIII. The first step in this study was the determination of the blade forces and their effect on the entire pintle structure. The method of determining blade forces was similar to that followed in the redesign of the blades described later in this chapter. The calculation of yawing and pitching moments from the blade forces was straightforward. Fig. 92 shows schematically the blade forces, the hinge forces and the yawing and pitching moments. In the analysis of the factors affecting the oscillating forces, the wind conditions were limited to certain typical cases.

The magnitude of the following factors was varied one at a time and the effect on each of the others was determined:

a. The angle of tilt (γ).
b. The distance from the pintle axis to the center of rotation (L).
c. The distance from the axis to the hinge-pins (e).
d. The weight of the rotating parts (W).
e. The speed of rotation (ω).
f. The mechanical coning-damping constant (β).
g. The number of blades (B).

In this part of the study only one wind condition was considered. The wind velocity selected—30 miles per hour—gives approximately the rated output of the turbine. The distribution of wind velocity vertically across the disc—the wind gradient—was assumed to be linear and such as to produce the maximum amplitude of periodic coning which was observed on the test unit.

Figs. 93–96 and 97–98 show the effect of varying the quantities b, c, and f listed above. Based on these results, two units were selected for further study, one with two blades, and the other with three. The three-bladed turbine was considered at this time because it was thought by some that it would be superior to a two-bladed unit, independently of the values selected for the variables listed above.

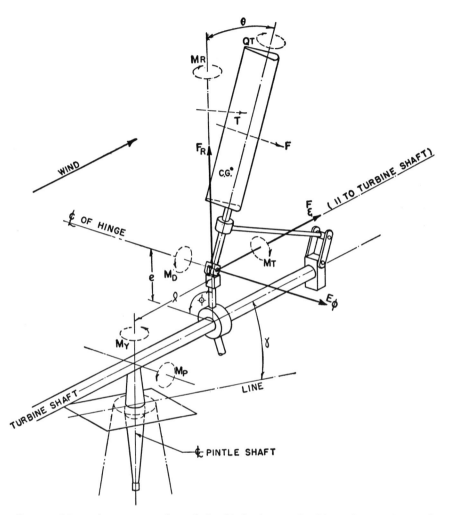

FIG. 92. Schematic representation of the blade forces, the hinge forces, the yawing moments, and the pitching moments.

Note particularly the dimension e, the distance between the hinge line and the turbine shaft.

In the test unit e had the value of 32.5 inches, which was the cause of some roughness in operation. In future units e will equal 0.0 inch.

159

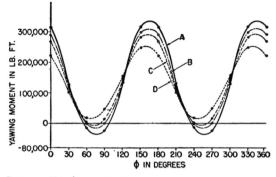

FIG. 93. Yawing moment.

Influence of the length L of Fig. 92.

Curve	A	L equals 12.00 feet
Curve	B	L equals 8.71 feet
Curve	C	L equals 6.00 feet
Curve	D	L equals 0.00 feet

Computed for a blade angle of 0 degrees in a wind velocity of 30 miles per hour with an assumed vertical velocity gradient across the turbine disc such that coning will amount to 5 degrees.

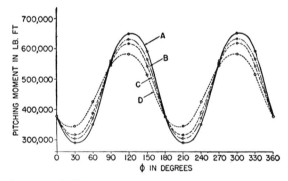

FIG. 94. Pitching moment.

Influence of the length L of Fig. 92.

Curve	A	L equals 12.00 feet
Curve	B	L equals 8.71 feet
Curve	C	L equals 6.00 feet
Curve	D	L equals 0.00 feet

Computed for a blade angle of 0 degrees in a wind velocity of 30 miles per hour with an assumed vertical velocity gradient across the turbine disc such that coning will amount to 5 degrees.

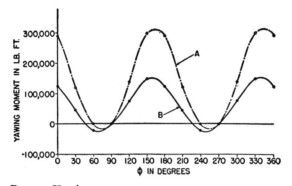

FIG. 95. Yawing moment.
Influence of the length *e* of Fig. 92.

Curve A	*e* equals 2.71 feet
Curve B	*e* equals 0.00 feet

Computed for a blade angle of 0 degrees in a wind velocity of 30 miles per hour with an assumed vertical velocity gradient across the turbine disc such that coning will amount to 5 degrees.

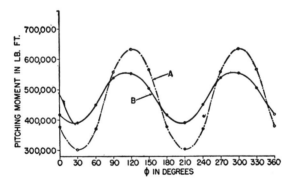

FIG. 96. Pitching moment.
Influence of the length *e* of Fig. 92.

Curve A	*e* equals 2.71 feet
Curve B	*e* equals 0.00 feet

Computed for a blade angle of 0 degrees in a wind velocity of 30 miles per hour with an assumed vertical velocity gradient across the turbine disc such that coning will amount to 5 degrees.

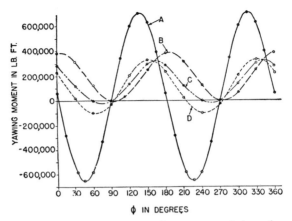

FIG. 97. Influence on the yawing moment of the value of the mechanical coning damping constant, β.

Curve A	$\beta = \infty$
Curve B	$\beta = 0.0$
Curve C	$\beta = 0.5$
Curve D	$\beta = 1.0$

These curves are computed for a blade angle of o degrees and a wind velocity of 30 miles per hour with an assumed vertical wind velocity gradient across the disc such that the coning angle will amount to 5 degrees.

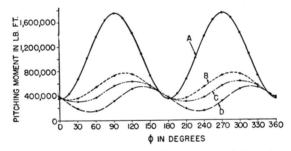

FIG. 98. Influence on the pitching moment of the value of the mechanical coning damping constant, β.

Curve A	$\beta = \infty$
Curve B	$\beta = 1.0$
Curve C	$\beta = 0.5$
Curve D	$\beta = 0.0$

These curves are computed for a blade angle of o degrees and a wind velocity of 30 miles per hour with an assumed vertical wind velocity gradient across the disc such that the coning angle will amount to 5 degrees.

162

In comparing these two units, three conditions of wind, blade angle, and vertical gradient of wind velocity were assumed. These three conditions were:

1. Wind 30 miles per hour; design blade angle; gradient to give maximum coning.
2. Wind 30 miles per hour; design blade angle; gradient to give average coning.
3. Wind 60 miles per hour; design blade angle + 24 degrees; gradient to give maximum coning.

The comparison between two- and three-bladed wind-turbines can best be shown by Figs. 99 to 104 on which are plotted the variations, respectively, of the yawing and pitching moments with variation in the blade position. It was found that this three-bladed unit, when exposed to a linear wind gradient, was not the best from the standpoint of yawing and pitching moments. The best three-bladed unit for a linear wind gradient would have fixed blades, i.e., no coning, and there would be no periodic yawing or pitching moment associated with such a unit. However, for a non-linear wind condition, such a unit would have large periodic yawing and pitching moments. Since the wind gradient usually is non-linear, it was concluded that the three-bladed wind-turbine holds no dynamic advantage over a well-designed two-bladed unit which, as was indicated previously, was more economical.

The greatest single improvement in a two-bladed unit, as compared with the design of the test unit, results from reducing the hinge distance e to zero. If circumstances had permitted, further studies would have been made, based on the assumption of zero hinge distance, in order to determine more accurately the effect of such things as tilt angle, blade weight, rotational speed, and coning-damping.

However, we believe that the results of such further studies would not have changed the conclusions.

Loadings for the Preproduction Model

Having fixed all of the dimensions under the control of the designer (other than member sizes), the next step was to compute loadings for an assumed mass distribution, beginning with the blades (Ref. 40-B). Since the dimensions of the preproduction unit were similar to those of the test unit, the mass distribution of the rotating parts was assumed to be the same as that of the test unit.

The distribution of mass along the blade axis from the root to the tip was assumed to be that of the test unit. The possibility of ice formation on the blades was allowed for by adding a constant mass per foot of blade length. The maximum ice load was that specified by von Kármán for the test unit—132 pounds per linear foot. (For an intermediate condition 25 per cent of this ice loading was used.)

The conditions for which loadings should be computed are not clear-cut in the case of wind-turbines, because we lack knowledge of wind behavior. The condi-

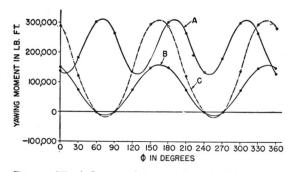

FIG. 99. The influence of the number of blades on the yawing moment. The conditions are: wind, 30 miles per hour; blades set at design blade angle; wind velocity gradient across the disc is taken to be that which will produce maximum coning.

    A. A three-bladed unit
    B. The two-bladed optimum unit
    C. The two-bladed test unit

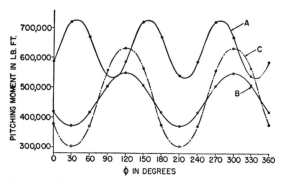

FIG. 100. The influence of the number of blades on the pitching moment. The conditions are: wind, 30 miles per hour; blades set at design blade angle; wind velocity gradient across the disc is taken to be that which will produce maximum coning.

    A. A three-bladed unit
    B. The two-bladed optimum unit
    C. The two-bladed test unit

164

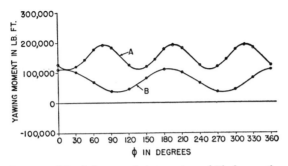

FIG. 101. The influence of the number of blades on the yawing moment. The conditions are: wind, 30 miles per hour; blades set at design blade angle; wind velocity gradient across the disc is taken to be **that** which will produce average coning.

A. A three-bladed unit
B. The two-bladed optimum **unit**

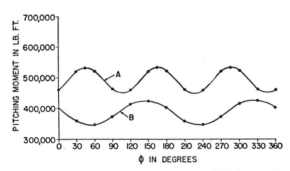

FIG. 102. The influence of the number of blades on the pitching moment. The conditions are: wind, 30 miles per hour; blades set at design blade angle; wind velocity gradient across the disc is taken to be that which will produce average coning.

A. A three-bladed unit
B. The two-bladed optimum **unit**

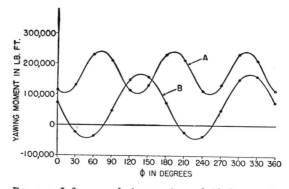

FIG. 103. Influence of the number of blades on the yawing moment. The conditions are: wind, 60 miles per hour; blades set at design blade angle plus 24 degrees; wind velocity gradient across the disc is taken to be that which will produce maximum coning.

    A. A three-bladed unit
    B. The two-bladed optimum unit

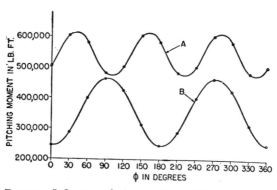

FIG. 104. Influence of the number of blades on the pitching moment. The conditions are: wind 60 miles per hour; blades set at design blade angle plus 24 degrees; wind velocity gradient across the disc is taken to be that which will produce maximum coning.

    A. A three-bladed unit
    B. The two-bladed optimum unit

tions investigated were set up as a result of the accumulated knowledge gained in operating the test unit, and in carrying out the special wind-research programs (Ref. 41-B).

These conditions fell into two general classes, operating and non-operating, which were further sub-divided:

   I. Operating conditions
      A. Fatigue loading conditions
         1. Constant velocity condition
         2. Varying velocity condition
      B. Maximum loading conditions
  II. Non-operating conditions
      A. Idling conditions
      B. Locked conditions

The forces on the blades—both mass and aerodynamic—can be expressed in terms of certain known coefficients and the unknown motions of the system.

Thus, it is possible to write differential equations of the motion of the system. The solution of these equations yields the unknown motion components, that is, displacements, velocities, and accelerations. These quantities, when multiplied by the proper coefficients, become the values of the aerodynamic and mass forces.

In addition to gravity forces, mass forces associated with the following accelerations were computed:

  1. The rotational centrifugal acceleration, which acts radially inward along a line perpendicular to the plane passing through the axis of turbine rotation.
  2. The rotational tangential acceleration, which acts along a line perpendicular to the plane containing the mass and the axis of turbine rotation.
  3. The centrifugal acceleration of coning, which acts along a line perpendicular to and passing through the coning axis.
  4. The tangential acceleration of coning, which acts along a line perpendicular to the plane containing the mass and the coning axis.
  5. The coriolis acceleration, which acts along a line parallel to the coning axis.
  6. The pitching acceleration, which acts along a line perpendicular to the plane containing the mass and the pitching axis of the blade.

The forces associated with the above accelerations are given by the products of the respective masses and their accelerations. Fig. 105 shows these forces in relation to the geometry of the turbine.

The aerodynamic forces are obtained in the form of lift, drag, and moment components. These are computed at a sufficient number of stations along the blade to permit smooth curves to be drawn. Values at other stations were obtained from the curves by interpolation.

All the elemental forces are resolved into three components acting respectively: parallel to the blade axis; perpendicular to the blade axis and perpendicular to the coning axis; and perpendicular to the blade axis and parallel to the coning axis.

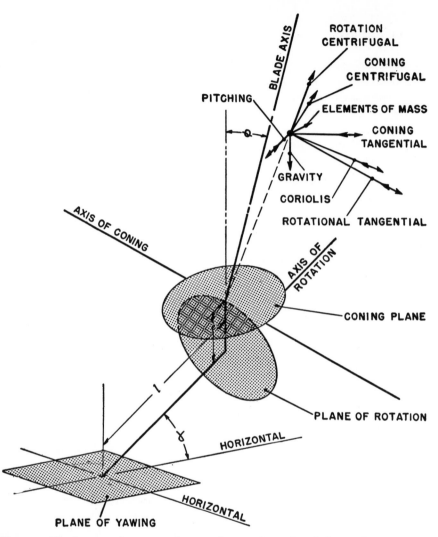

FIG. 105. The forces acting on an element of mass, shown in relation to the geometry of the turbine.

The forces were integrated to give the following quantities at each cross section of the blade:

A. Total force, parallel to the blade axis.
B. Shear and bending moment in the coning plane.
C. Shear and bending moment in the plane perpendicular to the coning plane.
D. Torque about the blade axis.

The four quantities provided the data for blade design.

The integrated forces at the root of the blade plus the integrated forces on the other parts were used for the design of the remainder of the structure. In so far as loadings for the rotating parts of the structure were concerned, the first cycle was completed. Loads for the remainder of the structure were obtained in a more approximate fashion. This was because information was urgently needed for rough layouts and for cost estimates.

Preliminary Design Studies and Layouts

Several types of blade were considered for the production unit. The Budd Company, builders of the original blades, designed new blades of two different

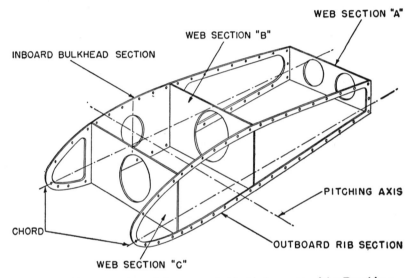

Fig. 106. Frame section of the stressed skin blade proposed by Dornbirer.

types. One was very similar to the original design, but with a modified and strengthened root section. The other was a flexible-skin blade in which all loads were carried by the spar. The skin of this blade was in sections approximately 6 feet long with no provision to transfer load from one section to the next. Both of these blades would have stainless-steel skin and ribs on an alloy steel spar. The spar would have a bolted, flanged connection to the shank. In these blades twist can be built in at no extra cost.

A stressed-skin blade which could be fabricated in short sections and field-assembled was designed by S. D. Dornbirer, Chief Construction Engineer, Wind Turbine Division of the S. Morgan Smith Company (Fig. 106). This design was originally an annealed stainless-steel skin on a carbon steel foundation. Annealed stainless is very little stronger than some of the carbon steels. For a weight penalty of 10 per cent, we could reduce the cost of material 25 per cent. The 100 per cent carbon steel construction would make the entire fabrication and assembly job much easier.

A mock-up of several sections was made in the S. Morgan Smith shops to aid in completing the design. Fig. 107 shows a view of this mock-up. If this type of blade were used, it would be built with no twist, since there is a distinct gain in fabrication by so doing. Sufficient investigation of this blade was made to indicate that it held promise.

Plywood blades for a 200-foot diameter wind-turbine were designed by the United States Plywood Corporation (Ref. 42-B). These blades were designed prior to the determination of the loadings to which the stainless and mild steel blades were designed. These blades are not, therefore, directly comparable to the steel blades. However, certain general comparisons between plywood blades and

FIG. 107. Mock-up of the section of the frame of the stressed skin blade proposed by Dornbirer.

steel blades can be drawn. The initial cost of plywood blades is probably less in both material and labor than that of metal blades. The weight is about the same. The efficiency should be slightly better, at least for the first few years.

Maintenance costs over the 20-year life would be much higher for the plywood blades. In fact, we do not know how the plywood blades would behave over such a long period when exposed to high winds, ice, oil, dirt, and other hazards. With the present knowledge of wood glues and surface protection, it is probably a little optimistic to say that plywood blades will stay in operation for 20 years.

Blades fabricated of an aluminum alloy were briefly discussed with several groups interested in that material. Preliminary studies were started with the Aluminum Laboratories Ltd. of Montreal, Canada, but were suspended when S. Morgan Smith abandoned the project.

Preliminary layouts indicated the following design features for the preproduction model, as shown in Fig. 108.

1. A redesigned A-frame with a tubular member and side tie rods. The name A-frame is no longer applicable.
2. The coning hinges are co-axial about the center line, thereby reducing e to zero (Chapter VIII).
3. A redesigned pitching mechanism eliminates universal joints.

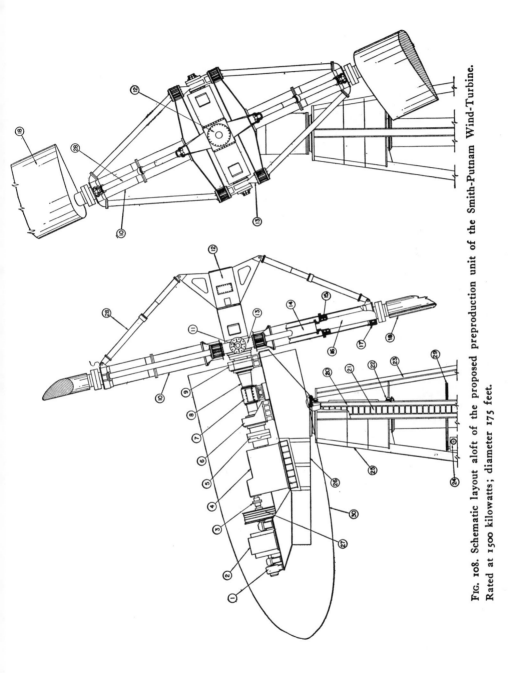

Fig. 108. Schematic layout aloft of the proposed preproduction unit of the Smith-Putnam Wind-Turbine. Rated at 1500 kilowatts; diameter 175 feet.

4. Oleo strut type of coning-damping mechanism.
5. A entirely welded hub-post bolted to a flange on the main shaft.
6. All welded pintle girder.
7. Hollow tubular pintle shaft.
8. Streamlined house and spinner.
9. All-welded tower.

The mechanical features of the turbine have been changed but little. Most bearings are anti-friction-ball, straight roller, or spherical roller.

The gear box incorporates a three-speed step-up rather than two as in the test unit. Minor simplifications are incorporated in this layout as a result of experience with the test unit, primarily to add to the ease of erection and maintenance.

The control problem is simplified by the substitution of an electric coupling for the hydraulic coupling. The generator would be phased on the line manually and left on until an emergency required taking it off. When the wind was too low to generate power, the coupling would be de-energized and the generator would run as a synchronous condenser under automatic voltage control. Several devices have been suggested to energize and de-energize the coupling at that wind velocity when the output becomes zero. With the full complement of safety devices, the turbine would operate as an automatic substation.

Summary

The objectives of the 1945 redesign of the Smith-Putnam Wind-Turbine were to provide: a structurally safe unit; a smooth operating unit; an economically competitive unit.

The first two objectives have probably been met, but the 1945 design, which weighed as much per kilowatt as the 1940 test unit, failed, as will be described in Chapter XI, to reach the third objective. Chapter XII describes speculations on how to reach this economic objective.

Chapter XI

ECONOMICS OF LARGE–SCALE WIND–POWER

Introduction

The preproduction unit, whose design is described in Chapter X, and the model of which is shown in Fig. 109, is a simplified and cleaned-up version of the test unit of 1940. It does not incorporate the modifications, still less the radical suggestions, put forward in Chapter XII.

This was because the engineers of the S. Morgan Smith Company had acquired six years of familiarity with the test unit. They felt confident that they could now go into production with a design of this general character and size. Furthermore, this familiarity provided the only basis for cost studies. To attempt to estimate closely the production costs of a design whose stresses had not been analyzed would obviously be a waste of time.

However, a comparison of Fig. 1 with Fig. 108 will show that, although the preproduction design was fundamentally identical with that of the test unit, it had been sufficiently modified in detail to require new shop drawings, which it was not feasible to prepare in the summer of 1945. Without shop drawings it was impossible to establish the production costs with finality. The methods by which we approximated the cost estimates are described in detail with respect to each, item.

Installed Costs

This 1945 cost study (Ref. 43-B) is based on a hypothetical production run of twenty 1500-kilowatt units, of which six would be installed on Lincoln Ridge in Vermont, while the other fourteen would be installed at unknown sites, the access costs of which were arbitrarily assumed.

Engineering Costs.

The man-hours spent in designing the test unit in 1939–1941, and in carrying out preliminary engineering studies of the preproduction unit in 1944 and 1945, formed the basis on which were estimated the costs of completely engineering a preproduction unit. At a conference in York, Pennsylvania, attended by the Chief Engineer of the S. Morgan Smith Company and most of the members of the Wind-Turbine Division, an engineering schedule was set up, on the assumption

FIG. 109. Scale model of the proposed preproduction unit of the Smith-Putnam Wind-Turbine.
Rated at 1500 kilowatts; diameter 175 feet.

that the design of the preproduction unit could be completed in twelve months. The engineering cost of such a design was estimated at $115,000, as itemized in Table XVI. An additional sum of $5000 per unit was allocated to cover the cost of design and production changes, and field engineering. For the twenty units, then, the total engineering cost would be $210,000, or $10,500 per unit.

Table XVI. breakdown of engineering cost

		Per Month	Per Year
1 Chief Engineer	at	$800.00	$ 9,600.00
1 Assistant Chief Engineer	at	500.00	6,000.00
3 Principal Engineers	at	400.00	14,400.00
1 Construction Engineer	at	400.00	4,800.00
4 Designers	at	300.00	14,400.00
6 Draftsmen			
(maybe 12 for 6 months)	at	200.00	14,400.00
2 Stenographers	at	150.00	3,600.00
		Total Labor	$ 67,200.00
		Overhead (30%)	20,160.00
			$ 87,360.00
		Travel	7,200.00
		Consulting	15,000.00
		Furnishings	7,000.00
			$116,560.00

Total man hours excluding stenographers 32,000.

Add $5000 per unit after the first unit for design and production changes and to provide for engineering during erection and testing.

For the purpose of this estimate, we used the following:

	Total	Per Unit
1 Unit	$115,000	$115,000
5 Units	135,000	27,000
6 Units	140,000	23,333
20 Units	210,000	10,500

Manufacturing Costs.

It is convenient to distinguish between the costs of standard items now in production and the costs of those items peculiar to the Smith-Putnam Wind-Turbine.

To allocate handling and other charges, the standard production items were further classified into three groups: Group one—items shipped by the manufacturer directly to the turbine site; Group two—items shipped to the S. Morgan Smith Company at York, Pennsylvania, for shop assembly; Group three—items shipped to York for machining and fitting before shop assembly.

The handling charges on these three groups were set by the S. Morgan Smith Company at 2½ per cent, 5 per cent, and 10 per cent of the net purchase price, respectively.

175

Items in Standard Production

Synchronous Generator.

As a result of the fourth approximation of the most economical design, described in Chapter IX, the synchronous generator selected for the preproduction unit had been rated at 1500 kilowatts, 2300 volts, with a speed of 900 revolutions per minute. Quotations were received from the Electric Machinery Manufacturing Company, the General Electric Company, the Allis-Chalmers Manufacturing Company, and the Westinghouse Electric and Manufacturing Company.

Main Gears.

In 1943, when the third study was under way to determine the most economical generator speed and rating, the Falk Company of Milwaukee, Wisconsin, had sent their Chief Engineer, Walter Schmitter, to Rutland to explore the problem. Until the S. Morgan Smith Company abandoned the project on December 31, 1945, the Falk Company gave it close cooperation and made many design and cost studies to assist in the determination of the most economical capacity and speed of the generator and gear. When the results of the fourth determination in 1945, described in Chapter IX, indicated a 1-to-30 gear to drive the generator, the Falk Company designed a series of gear boxes for this duty. At one end of the series was an ultralight gear assembly of a type used in destroyers, and with a high pound price. At the other end of the series was a heavy assembly with a low pound price. Considering the effect of weight on the rest of the turbine structure, an intermediate design was selected.

Estimates of the gear assembly were also supplied by the General Electric Company and the Westinghouse Electric and Manufacturing Company.

The Electric Coupling.

The problem of providing a coupling between the generator and the gear box, to serve as a torque-limiting device and also to provide a speed change to which the Woodward Governor could respond, was explored by the Dynamatic Corporation of Kenosha, Wisconsin, and the Electric Machinery Manufacturing Company of Minneapolis, Minnesota.

Governor.

The governor specifications were explored with the Woodward Governor Company of Rockford, Illinois. Experience with the test unit indicated that the production governor could be far simpler than the one used on the test unit.

Bearings.

There were many anti-friction bearings in the Smith-Putnam Wind-Turbine, the largest being 48 inches in diameter, at the head of the pintle shaft (Fig. 108).

Among the most important bearings were the blade-shank bearings, the main-shaft bearings, and the pintle-shaft bearings. These bearings were either special, or at least nonproduction, items. Most of the other bearing requirements could be met by standard catalogue items.

The S. K. F. Industries and the Bantam Bearing Division of the Torrington Company cooperated closely with us in studying our bearing problems. Where necessary, new bearings were designed. After giving consideration to many alternative schemes, an assembly was selected which appeared to be the most economical.

Switchgear.

A functional specification, far simpler than in the case of the test unit, was prepared by the customer, the Central Vermont Public Service Corporation, whose engineers specified the switchgear necessary to discharge the required functions. Cost estimates were obtained from the Allis-Chalmers Company, the Westinghouse Electric and Manufacturing Company, the Electric Machinery Manufacturing Company, the General Electric Company, and the G & N Engineering Company.

Flexible Couplings.

The cost of the flexible couplings for the low-speed shaft was obtained from the Falk Company and for the high-speed shaft from the Farrell Birmingham Company.

Elevator.

S. Morgan Smith Company engineers estimated the cost of the production elevator, based on that of the test unit design.

Service Hoist.

Experience on the test unit had indicated that, while a service hoist was essential, it need not be elaborate. The cost of such a hoist was estimated.

Miscellaneous Electrical Equipment.

The cost of miscellaneous electric equipment including such items as slip rings, lighting circuits and fixtures, control circuits, and safety circuits, was estimated by S. Morgan Smith Company engineers.

The Tower.

Quotations were obtained for two different types of tower. The American Bridge Company of Pittsburgh, Pennsylvania, who had designed and built the four-legged tower for the test unit, supplied quotations for a similar four-legged preproduction tower.

The Chicago Bridge and Iron Company, of Chicago, Illinois, quoted costs of a shell-type tower, looking like a truncated cone, and built up of sheets of steel plate. Quotations were submitted for each of four design conditions. Two of the design conditions varied with the foundations to be encountered, and two with the buckling stresses to be allowed. It was not feasible to include in this study a determination of the actual foundation conditions on the summit of Lincoln Ridge, so we assumed what we thought would be the worst condition, that is, a crumbling schist, like that found on the summit of Grandpa's Knob. Also, we used the lower of the two buckling stresses, with the result that the estimated cost of the shell-type tower, which we used for comparison, was the highest of the four estimates supplied by the Chicago Bridge and Iron Company.

There was little to choose between this estimate for the shell tower and the estimate for the four-legged tower.

Paint.

We estimated the area to be painted at 30,000 square feet, including blades. E. I. DuPont DeNemours estimated $2.25 per 100 square feet for a four-coat spray application.

Items Peculiar to the Smith-Putnam Wind-Turbine

In this category are the blades, the hub assembly, and the pintle assembly. It must be emphasized that detailed shop drawings of the production design were not available. To make cost estimates of these components of the wind-turbine, it was necessary first to estimate the over-all weight of each component, and second, to apply to this weight a unit pound price based on experience with similar structures in the S. Morgan Smith Company shops.

The Blades.

Studies had indicated that mild-steel blades, with surfaces either painted or galvanized, or treated in some other way, to resist corrosion, would be somewhat more economical than stainless-steel blades. Various types of structure had been roughed out, from the nearly monocoque and heavily skin-stressed, to a spar and rib structure covered with a non-stressed skin. An intermediate type had been selected, which could be manufactured to advantage in the shops of the S. Morgan Smith Company and which looked as though it would reduce the cost of field erection. Based upon the new blade loadings described in Chapter X, the approximate thicknesses of the various members were determined. From this information the blade weight was estimated at 67,000 pounds for two blades.

The pound price of this blade structure was independently estimated by J. Oerter and W. H. Hollingshead of the S. Morgan Smith Company Estimating Department. The two estimates were in close agreement, and averaged $0.484 per pound F.O.B. York, Pennsylvania, in lots of one pair, and $0.440 per pound in lots of 20 pairs, equivalent to a cost of $29,480 for each pair, in lots of 20 pairs.

Hub Assembly.

This includes all special items between, but not including, the blades and the main shaft. The over-all weight of these items was estimated after considering two somewhat independent approaches:

A. The weight of all corresponding parts on the test unit was found to be 175,900 pounds, and, of this total, about 115,800 pounds consisted of parts which might reasonably be expected to go down in weight through redesign and the use of welding instead of riveting. Weight reduction from this cause was estimated at 17.5 per cent. Thus, the estimated weight of the preproduction hub assembly would be 155,600 pounds.

B. The weight of these parts estimated from Dornbirer's 1945 sketches for the preproduction unit was 136,620 pounds. Since these sketches were based on inadequate analysis, it was felt that the weights should be increased by an arbitrary factor of 10 per cent, yielding a weight estimate of 150,300 pounds, broken down approximately as follows:

Plate steel	50 per cent
Cast steel	25 per cent
Forged steel	15 per cent
Miscellaneous	10 per cent

The weight estimate adopted by the engineers was 155,000 pounds.

The pound price was again estimated independently by J. Oerter and W. H. Hollingshead. They compared the known costs of similar structures built by S. Morgan Smith Company, in which the proportions of plate steel, castings, and forgings were approximately the same as in the foregoing tabulation. Again, the two estimates were very close and averaged $0.310 per pound for single units and $0.277 per pound in lots of 20, giving a total cost for the hub assembly of $42,935.

Pintle Assembly.

This assembly includes all special items from, and including, the main shaft to, and including, the tower cap. It also includes the yaw-mechanism, although part of this may be located below the tower cap.

The weight of the pintle assembly was estimated by considering the evidence found in two more or less independent approaches:

A. The weight of all corresponding parts on the test unit was found to be 200,000 pounds.

B. Dornbirer's 1945 sketches for the preproduction unit showed weights itemized as follows:

Plate steel	175,000 pounds
Castings	20,000 pounds
Forgings	30,000 pounds

The engineers felt that the weight of the plate steel items should come down as a result of a more careful analysis and design. This item was, therefore, arbitrarily reduced to 150,000 pounds, giving a total weight of 200,000 pounds.

For estimating purposes the engineers accepted 200,000 pounds as the weight of the pintle assembly, itemized as follows:

Plate steel	73 per cent
Forged steel	15 per cent
Cast steel	10 per cent
Miscellaneous	2 per cent

The pound price was determined in the same manner as for the hub assembly, and averaged $0.260 per pound for a single unit and $0.243 per pound in lots of 20 units, making a total cost for the pintle assembly of $48,600 for each of 20 units.

Patterns, Tools, Jigs, and Fixtures.

The cost of patterns, tools, jigs, and fixtures was estimated by the respective departments of the S. Morgan Smith Company.

Installation Costs

Freight.

The Rutland Railroad estimated that the average freight rate from York, Pennsylvania, to Rutland, Vermont, would be $0.47 per 100 pounds.

The shipping weight of the unit was taken to be the weight of the items manufactured at York, Pennsylvania, plus the weight of the items shipped to the site via York, giving a total shipping weight of 437,000 pounds.

Although this freight rate applied only to the first 6 units, we arbitrarily applied it also to the remaining 14 units.

Land.

On the recommendation of the Central Vermont Public Service Corporation this item was estimated to cost nothing.

Road.

The cost of an access road is obviously variable. On paper we laid out a road which would serve 6 sites on Lincoln Ridge. We reconnoitered the ridge on foot and explored the problem with road-building contractors, whose lowest estimate was $44,760 for the 6 units, or $7460 per unit. We arbitrarily assumed that this cost would not be exceeded in the case of the remaining 14 units.

Erection.

Erection costs were estimated by S. D. Dornbirer, who had erected the test unit. He made a complete breakdown of the erection procedure, which he had greatly simplified. He estimated the erection cost for one unit on Lincoln Ridge as $57,000, and for 6 units as $192,000, or $32,000 per unit. It was assumed that some of the remaining 14 sites would be more accessible and subject to less rugged weather; and, accordingly, an average erection cost of $30,000 per unit was assumed for the 20 units.

Connection Cost—Transformers.

A study showed that for a block of 6 wind-turbines, the most economical arrangement was to use a single transformer bank and run lines at generator voltage to each turbine. The cost of such a transformer bank, with its associated equipment, lightning arresters, disconnects, etc., is approximately $22,000 or $3600 per unit.

High-Line.

The cost of the high-tension transmission line is an item which obviously will vary greatly with circumstances. At Grandpa's Knob it was necessary to build only 2.8 miles of transmission line. At Lincoln Ridge, on the other hand, 24 miles of transmission line would be needed. The Central Vermont Public Service Corporation estimated the cost of a 44-kv line at about $5000 per mile, making a total cost for Lincoln Ridge of $120,000 or $20,000 per unit. It was felt that Lincoln Ridge is probably as far from an existing power line as one would find in New England, or in many other places; and for this reason it was arbitrarily assumed that $15,000 would be the average cost of the transmission line for each of the 20 units.

Total Installed Cost

The total installed cost of the wind-turbine is summarized in Table XV. It is seen that the cost at the switchboard is $191.11 per kilowatt, and the cost at the point of connection to the existing high-line, including contingency, is $204.75 per kilowatt. The installed weight, including foundation steel, but excluding transformers and connecting line, is 497 pounds per kilowatt, and the average unit cost is $0.39 per pound.

Annual Charges

The annual charges are determined by multiplying the total investment by a percentage determined by a consideration of the following factors:

a. Interest rate on bonds.
b. Dividend rate on stocks.
c. Property tax rate.
d. Income tax rate.
e. Depreciation rate.
f. Operating costs.
g. Maintenance costs.

The proportionality between the amount of the investment put into bonds and that into stocks has a large effect on the total rate since current dividend rates are so much higher than current interest rates. All of the above listed factors will vary with different utilities.

In general, the annual charge which a private company would make for a wind-turbine installation would be in the range from 12 per cent to 15 per cent. Similar charges by a government agency would be in the range from 6 per cent to 10 per cent. An example of the computation for a specific utility will be given in the section on the cost of wind-power to the Central Vermont Public Service Corporation.

Annual Output

It is convenient to measure the annual output in terms of the number of kilowatt-hours put out in one year per kilowatt of rated generator capacity. Table I of Chapter I summarizes in these units the outputs to be expected on the windy islands of the world, using a scale which runs from "over 6000" kilowatt-hours per kilowatt per year to "less than 2000." At Grandpa's Knob we realized about 1200 kilowatt-hours per kilowatt per year. In 1945 we estimated that a battery of six 1500-kilowatt wind-turbines on Lincoln Ridge would generate an average of 3500 kilowatt-hours per kilowatt per year. Since then, wind-velocity measurements on the Horn of Mount Washington, where the deformation of the balsams is similar to that on the southern end of Lincoln Ridge, have confirmed our belief that the average output on Lincoln Ridge will be about 3500, while, in the case of a unit erected on Mount Abraham, the southernmost end of the ridge, the output would reach 4500 kilowatt-hours per kilowatt per year, or more.

Energy Costs

The cost of energy is computed by dividing the annual charges in dollars per kilowatt by the annual output in kilowatt-hours. The result, in dollars per kilowatt-hour, is converted to mills per kilowatt-hour by multiplying by 1000.

Fig. 110 is a nomogram showing the relationship between installed costs, annual charges, annual output, and the resultant energy cost at the switchboard at the foot of the tower.

Thus, if private capital installs a wind-power plant for $100 per kilowatt with annual charges of 12 per cent in a wind regime such that the output is 4000 kilowatt-hours per kilowatt per year, the energy cost will be 3 mills per kilowatt-hour. If, on the other hand, the installation had been made by a Government agency, with annual charges at 6 per cent, the energy cost would have been 1.5 mills per kilowatt-hour.

Components of the Worth of Wind-Power

The worth of wind-power to any system has four components: capacity value, reactive value, value because of predictability, and energy value. A study of the

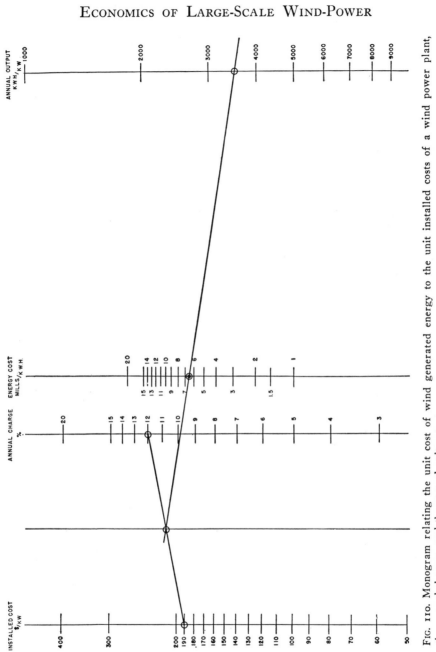

FIG. 110. Monogram relating the unit cost of wind generated energy to the unit installed costs of a wind power plant, the annual charges, and the annual unit output.

worth of these components was carried out by Jackson and Moreland, whose findings are summarized in the following sections.

Capacity Value.

A wind-power installation is not an isolated source of firm power. In order that it may create some capacity value, there must be other generating equipment available on the same system. Thus, if there is an excess of water-wheel capacity accompanied by adequate pondage on a certain system, the addition of a block of wind-power would create some additional firm capacity. Alternatively, if a hydro-electric system was already in balance with respect to its water-wheel capacity, it would be necessary to add some water-wheel capacity in order to firm up whatever wind capacity it was proposed to add.

The probable capacity value of wind-power as found on Lincoln Ridge would vary from $5 per kilowatt on a system with short hydro-storage periods (and high daily load-factors during the months of low wind-power) to $40 per kilowatt on systems with long hydro-storage periods (realizing thus the average wind-power) and modest annual load factors.

Of course the usable wind-energy may at times exceed the system load, while simultaneously the hydro system may be incapable of storing more water. In these circumstances, not all the wind-energy can be used to back off fuel-generated energy. Such excess wind-energy has no value.

No credit is given to wind-power for any capacity value arising from the diversity in output of many wind-power plants spread over a vast area. The worth of such capacity value would be exceeded by the transmission costs incurred in getting the wind-energy from the spots where the wind was blowing to the spots where it was not.

Reactive Value.

In certain types of wind-power plants the synchronous generator may be left on the line as a synchronous condenser when there is not enough wind for power generation. When acting as a condenser it will provide reactive kilovolt-amperes for the system. The worth of this will vary from $0.00 per kilowatt, when the system is adequately supplied with reactive power or when the wind-turbine is so located that the reactive power cannot be used, to $8.50 per kilowatt when the system is in need of all the reactive power that can be provided from idling wind-turbines and when these turbines are so located that all the wattless is useful for system purposes.

Predictability Value.

In Chapter V it was reported that it would be possible to predict 12 hours in advance that one of three conditions would prevail at a wind-power plant during a 6-hour period. The three conditions were:

1. No output—no winds of more than 17.5 miles per hour.
2. Some output—winds of more than 17.5 miles per hour.
3. Full output—no winds of less than 34 miles per hour.

It was reported that the results of several hundred such predictions, analyzed qualitatively, show that the correlation factors were high enough to be useful, averaging about 0.80 for the 12-hour prediction.

Under certain circumstances this should make it possible for a power dispatcher to reduce some of his "floating" and "banked" standby fuel generating capacity. The economic value of these predictions would tend to offset any increase in the cost of dispatching which might result from the fluctuating nature of wind-energy. An analysis of individual systems might show that in certain cases this degree of predictability would endow wind-power with a small additional capacity value.

Energy Value.

The worth of wind-energy is measured by the extent to which a block of wind-power reduces the cost of energy which must otherwise be generated on the interconnected system. In a steam or a Diesel power plant the cost of generating energy is chiefly the cost of fuel, with some small additional charges for lubricants, supplies, and maintenance resulting from the actual output of energy from the plant, as distinguished from similar expenses which are fixed and continue whether the plant is operating or not.

Such energy costs in 1945 typically ranged from 2.5 mills per kilowatt-hour at a large, modern steam station, to 6 mills per kilowatt-hour at a small, older station, and 10 mills per kilowatt-hour in the case of a small isolated Diesel plant.

Table XVII shows typical 1945 costs of fuel generation on systems capable of absorbing blocks of wind-power varying in size from 100,000 kilowatts to 100 kilowatts.

Evaluation of Wind-Power.

The components of the worth of wind-power vary in value depending on whether the wind-power installation stands alone or is backed up by hydro, or by fuel, or by combined hydro and fuel (Ref. 44).

Wind-Power Alone.

Where wind-power is backed up by other power sources, its applicability is as broad as the applicability of electric power generally. Where back-up from other sources is not feasible, the applicability of wind-power is limited to those uses which can practicably or economically employ intermittent and variable power.

Practicability requires that operation of the process or service can be reduced in rate, or interrupted, without damage to equipment or disorganization of the process and without undue inconvenience or economic loss. Processes requiring much time and expense for start-up after an interruption, or involving loss of

TABLE XVII. VALUE OF WIND-ENERGY UNDER VARIOUS ASSUMPTIONS

Size of Block of Wind-Power	Kwh. of Wind per Year		Typical 1945 Costs of Fuel-Generated Energy		Capitalized Value per Kw. of Wind Generator at				
	Available in Wind	Assumed Usable in Load	Mils/Kwh.	$/yr.	15%	12%	10%	8%	6%
Very large blocks, say, *100,000 kw.*	2500	2200	2.5	$ 5.50	36.67	45.83	55.00	68.75	91.67
	3000	2700		6.75	45.00	56.25	67.50	84.38	112.50
	3500	3200		8.00	53.33	66.67	80.00	100.00	133.33
	4000	3700		9.25	61.67	77.08	92.50	115.63	154.17
	4500	4200		10.50	70.00	87.50	105.00	131.25	175.00
	5000	4700		11.75	78.33	97.92	117.50	146.88	195.83
Large blocks, say, *50,000 kw.*	2500	2300	3.0	6.90	46.00	57.50	69.00	86.25	115.00
	3000	2800		8.40	56.00	70.00	84.00	105.00	140.00
	3500	3300		9.90	66.00	82.50	99.00	123.75	165.00
	4000	3800		11.40	76.00	95.00	114.00	142.50	190.00
	4500	4300		12.90	86.00	107.50	129.00	161.25	215.00
	5000	4800		14.40	96.00	120.00	144.00	180.00	240.00
Medium blocks, say, *10,000 kw.*	2500	2400	3.5	8.40	56.00	70.00	84.00	105.00	140.00
	3000	2900		10.15	67.67	84.58	101.50	126.88	167.17
	3500	3400		11.90	79.33	99.17	119.00	148.75	198.33
	4000	3900		13.65	91.00	113.75	136.50	170.63	227.50
	4500	4400		15.40	102.67	128.33	154.00	192.50	256.67
	5000	4900		17.15	114.33	142.91	171.50	214.38	285.83
Small blocks, say, *1000 kw.*	2500	2500	5.0	12.50	83.33	104.17	125.00	156.25	208.33
	3000	3000		15.00	100.00	125.00	150.00	187.50	250.00
	3500	3500		17.50	116.67	145.83	175.00	218.75	291.67
	4000	4000		20.00	133.33	166.67	200.00	250.00	333.33
	4500	4500		22.50	150.00	187.50	225.00	281.25	375.00
	5000	5000		25.00	166.67	208.33	250.00	312.50	416.67
Very small blocks, say, *100 kw.*	2500	2500	8.0	20.00	133.33	166.67	200.00	250.00	333.33
	3000	3000		24.00	160.00	200.00	240.00	300.00	400.00
	3500	3500		28.00	186.67	233.33	280.00	350.00	466.67
	4000	4000		32.00	213.33	266.67	320.00	400.00	533.33
	4500	4500		36.00	240.00	300.00	360.00	450.00	600.00
	5000	5000		40.00	266.67	333.33	400.00	500.00	666.67

expensive material due to interruption, are not suitable. Processes requiring heavy capital investments or heavy fixed operating costs in relation to the value of power would probably be uneconomic since variable and intermittent power reduces the use-factor of the investment and organization.

The use of intermittent power requires either availability of operators to start up equipment which may have been shut down and to regulate its operating rate in accordance with the variable power supply; or it requires investment in automatic controls to perform the same functions. Expense for maintaining operators in idleness, to start up processes which have been shut down, may be mitigated in some instances if the duration of the idle period and its termination may be predicted with some degree of reliability, thus making it possible to achieve a degree of coincidence between the normal time-off for operation—or assignment to other duties—and the period of deficient wind.

Listed below are some types of power application which, under favorable conditions, might profitably employ variable and intermittent power.

1. Pumping
    a. Pumping water to reservoirs for industrial and municipal supply purposes.
    b. Pumping water for irrigation and land reclamation.
    c. Pumping water into salt deposits for brine production or pumping up brine solutions.
2. Electrolytic deposition of metals
    a. Refining electrolytic copper from blister.
    b. Manufacture of electrolytic lead.
    c. Manufacture of electrolytic powdered iron for powdered iron metallurgy, etc.
3. Inorganic electrolyses: Electrolysis of water for production of hydrogen and oxygen (which could be compressed into cylinders by use of the same power source).
4. Mechanical power uses
    a. Manufacture of ice in standard equipment for isolated communities.
    b. Manufacture of distilled water by the vapor compression process for isolated communities having no regular fresh water supply, such as islands.
    c. Compression of gases for storage, either compressed or liquefied, as at the Cleveland liquefied natural gas storage plant.
5. Untended airway beacons, as in the Arctic or the great deserts.

It is obviously impossible to generalize about the worth of wind-power in these applications. The worth would vary with each case, and could be determined only by a special study.

Wind-Power with a Hydro System.

In conjunction with a simple hydro system without fuel auxiliary, and on which there exists a water-wheel capacity for supplying secondary unfirmed power, the addition of wind, where adequate pondage is available, would firm such surplus capacity to approximately the extent of the average rate of wind generation during the low water season.

If no such surplus exists the same additional firm power can be created by the simple addition of water-wheel generators in the amount of the average wind generation.

There are regions where the costs of hydro development are such that auxiliary fuel plants are not economical. In some of these regions, all the power necessary in the foreseeable future can be generated by hydro developments under these same economic conditions. In this case, it is not likely that wind-power can be justified.

The use of wind-power with pumped storage in combination with an existing hydro system will parallel the foregoing case in all respects except when the pumped storage can be so located that its water can be discharged through an existing hydro plant. In such a case, if excess wheel capacity is available in the existing plant, the benefit will obviously be the cost of this wheel capacity plus the costs for additional pumping and storage facilities due to the increased demands on the storage occasioned by the wind.

Wind-Power with a Fuel System.

When used in conjunction with fuel-generated power, with no hydro capacity, where the capacity of the steam or Diesel bears a normal relation to the system load, it seems reasonable to say that a wind-power plant can create no capacity value, for it will not reduce the size of the steam plant required to carry the load when there is no wind.

However, the wind-energy will displace the fuel-generated energy; and the worth of the wind-energy may be measured by the incremental cost of the fuel-generated energy so displaced. If, for example, the wind-power installation generates 3000 kilowatt-hours per year per kilowatt of wind-generated capacity, and if the incremental cost of the fuel-generated energy is 3 mills per kilowatt-hour, then the wind-energy would be worth $9 per kilowatt per year. It is understood that the rate of wind generation is here assumed to be never in excess of the difference between the system load and the minimum practicable rate of generation from the fuel plant.

Table XVII shows typical costs of generating energy in steam and Diesel power plants of various sizes and at various fuel prices. It will be seen that the cost of fuel-generated energy goes down as the scale of the operation increases, thus reducing the value of wind-energy as the block of wind-power increases in size.

Wind-Power with a Combined Fuel and Hydro System.

The energy value of wind-power to a combined water and fuel system is the same as on a simple fuel system, provided the fuel plants are similar in both cases.

Wind-power, however, added to such a system, will frequently also have a capacity value when backed up by the necessary new or existing water-wheel capacity and pondage. Where physical conditions are favorable, some of the wind-energy may be used to reduce the draw-down from the storage reservoirs while the wind is blowing, using this increment of stored water in the surplus hydro-generating capacity during hours when the wind is deficient, thereby increasing the firm capacity of the existing surplus water-wheels; or even making it economic to add further water-wheel capacity. As a result, the requirement for fuel capacity is decreased and the investment in additional fuel plants is avoided.

The addition of wind-power to a combined stored water and fuel system can create capacity value equal to the average wind-power generation during the period of draw-down at the storage reservoirs. The worth of this capacity value will be the saving in investment cost in providing this amount of generating capacity less the cost of providing water-wheel capacity in an equal amount. It is normal for hydro systems to have water-wheel capacity in excess of the firm hydro value and on such systems the immediate saving by the installation of wind-power is the gross value of the kilowatts created without such a deduction for installing an equal amount of water-wheel capacity. It should be recognized that this gross value, in contrast with the net value, will persist only so long as the water-wheel capacity would have been in excess of the firm capacity on the system load. For, with a growing system load, as the peak increases, the need for water-wheel capacity per unit of flow in the river increases.

Of course, as in the case of the use of wind-power on a simple fuel system, if the load at any time drops to a value as low as the minimum practicable steam plus hydro-generation, then the energy value of the wind-power during such hours of operation would be lost.

The Worth of Wind-Power to the Central Vermont Public Service Corporation

In October, 1945, the worth to the Central Vermont Public Service Corporation of a block of 9000 kilowatts of wind-energy installed on Lincoln Ridge was determined by Jackson and Moreland (Ref. 45-B). This study was made in collaboration with the Utility Company, but the conclusions have not been reviewed or approved by them. However, Jackson and Moreland believe that they have made the best estimate of the worth of wind-power to this Utility Company that could be made in the time available.

The analysis was confined to the main Central Vermont Public Service Corporation system comprising plants located on Otter Creek and its tributaries. The plants in this group are interconnected electrically and there are connections with Bellows Falls Hydro-Electric Corporation for purchased power which supplies about half of the system's total requirement. Isolated properties of Central Vermont Public Service Corporation were excluded.

Some eighteen hydro stations varying in capacity from 100 to 3400 kilowatts comprise this system, whose full load generating capacity totals some 19,000 kilowatts. For purposes of this study the plants were considered in three groups. The first group, totaling 11,200 kilowatts, has large storage reservoirs more than capable of handling the variations in stream-flow which occur within a period of a year, and with some capacity, although limited, to handle differences in annual flow as between one year and the next. The second group comprises two plants totaling 3900 kilowatts, and has a small reservoir capacity considered to be adequate to handle variations in load shape for a period of a week or two but with insufficient capacity to handle seasonal fluctuations in flow. The third group comprises a considerable number of small run-of-river plants, also totaling 3900 kilowatts, but with no storage and with ponds capable of maintaining rated output in dry weather for very short periods—in some cases not at all.

The system has no steam plants of its own, but the interconnection with Bellows Falls Hydro-Electric Corporation, which in turn interconnects with large steam plants, reflects the costs of steam-generated power. The contract provides a demand charge of $1.25 for each kilowatt of monthly demand and an energy charge of $0.009 per kilowatt-hour for the first 250 hours use of the demand, and $0.005 on additional energy with a discount of $0.001 on all kilowatt-hours. There is a coal charge which, at the time of the study, just about offset the discount. The demand charge is based upon the demand in certain peak hours of the heavy load season, as defined in the contract, with provisions to permit exceeding this demand in off-peak hours and seasons up to certain limits. For the purposes of this study the contract terms are the equivalent in cost to $0.005 per kilowatt-hour of energy and $27 per year per kilowatt of demand, as defined in the contract. To broaden the analysis, a second set of calculations was made assuming a situation in which the energy charge would be $0.0045 and the demand charge $15 per year per kilowatt.

Fixed Charges and Operating Costs

For the block of capacity on Lincoln Ridge, Jackson and Moreland, in conference with Central Vermont Public Service Corporation, agreed upon the fixed charges and operating costs in Table XVIII. The 3.36 per cent for depreciation of the wind-power development and the transmission lines corresponds to twenty years of life, with reinvestment in a sinking fund at 4 per cent.

Similarly, the 1.78 per cent for the depreciation of the hydro plant corresponds to a life of thirty years.

In the absence of actual wind velocity measurements on the summit of Lincoln Ridge, it was necessary to assume a wind regime by interpolating in the meteorological and ecological evidence found in New England on higher and lower summits. It was recognized that the outputs of the six units on the ridge would vary considerably among themselves, as indicated in Table IX, but it was assumed that the average output of the block would be not less than 3500 kilowatt-hours

TABLE XVIII. ITEMIZATION OF ANNUAL CHARGES

	Wind-Power Development	Transmission Lines	Hydro-Plant Additions
Interest 40% bonds at 3.25%	1.30%	1.30%	1.30%
Dividends 60% stock at 6.50%	3.90	3.90	3.90
Property Tax	2.00	2.00	2.00
Federal Income Tax at 25%	1.30	1.30	1.30
Depreciation	3.36	3.36	1.78
	11.86	11.86	10.28
Operation and Maintenance	$ 2.50/kw-yr	1.00	1.00
		12.86	11.28

per year per kilowatt of wind-generated capacity, of which there would be a net delivery to the Central Vermont Public Service Corporation system of 3320 kilowatt-hours after allowing for transformation and transmission losses.

The monthly distribution of this wind-energy was estimated by S. Morgan Smith Company to be as follows:

Month	Per Cent
January	11.3
February	10.5
March	11.0
April	7.5
May	6.8
June	6.3
July	5.0
August	5.8
September	5.7
October	8.5
November	9.6
December	12.0
Total	100.0

The energy value of the wind-power is measured by the worth of the resultant decrease in the quantity of purchased energy. The capacity value of the wind-power is measured by the resultant decrease in purchased demand, which depends upon the existence of hydro capacity to meet such demand when the wind does not blow. The estimates indicated that about 3600 kilowatts of additional hydro capacity would be required on the Central Vermont Public Service Corporation System to firm the 9000 kilowatts of wind. This is about 0.4 kilowatt of hydro per kilowatt of wind.

Against this background of physical plant, contractural obligations, and assumed wind velocities, Jackson and Moreland studied the rated and actual capacities of the hydraulic turbines on the system, the predictions of load growth, and the estimates of operating and maintenance costs of wind-power, as supplied by the Central Vermont Public Service Corporation. To arrive at a simplified dispatch, tentative agreements were reached with the Utility engineers as regards typical load shapes; reservoir operation and grouping of water flows into seasons; and grouping of stations into storage, semi-storage and strictly run-of-river. Agreement was reached as to the locations of the future development of the 3600 kilowatts of hydroelectric power required to firm the 9000 kilowatts of wind-power.

Based on this information, the most economical dispatch of the generated capacity, with and without wind, was worked out. This was held to be a sound approach since it was felt that corrections for departure in practice from these idealized operations would tend to cancel out. Finally, Jackson and Moreland estimated that the average cost of adding the additional 3600 kilowatts of hydro capacity would be $130 per kilowatt.

The economics of adding wind-power to the system of the Central Vermont Public Service Corporation is summarized in Tables XIX, XX, XXI, and XXII, the first two showing the value of wind-power based only on savings in energy, and the last two including savings in the demand charges. Table XXIII is a comparative summary of the foregoing tables. The last line represents the worth of the wind-power development and transmission line, including an additional credit for reactive kilowatts of $4 per kilowatt of installed capacity of wind-turbines.

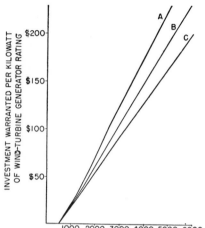

FIG. 111. Variation in the worth of wind-power to the Central Vermont Public Service Corporation with variations in annual output.

Curve A   Charges for power purchased from other systems:
    Energy Charge—$0.0050 per kilowatt-hour
    Demand Charge—$27.00 per kilowatt-year
Curve B   Most likely value
Curve C   Charges for power purchased from other systems:
    Energy Charge—$0.0047 per kilowatt-hour
    Demand Charge—$15.00 per kilowatt-year

On Fig. 111 are plotted the variations in the worth of wind-power to Central Vermont Public Service Corporation as the annual output varies from 2000 to 6000 kilowatt-hours per kilowatt. The high, low, and most likely values are plotted.

TABLE XIX. COMPARATIVE FINANCIAL STATEMENTS FOR 9000 KILOWATT WIND-POWER DEVELOPMENT
(Assuming Purchased Energy at $0.0050/kwh.: No State Tax on Wind-Energy; No Associated Hydro Plant for Reducing Purchased Demand)

BALANCE SHEET

Assets	Without Wind	With Wind	Net	Liabilities	Net
Hydro Plant Additions	—	$ —	$ —	Bonds (40%)	$ 424,300
Wind-Power Development	—	950,700	950,700	Stock (60%)	636,400
Transmission Line	—	110,000	110,000		
	—	$1,060,700	$1,060,700		$1,060,700

INCOME STATEMENT

Revenue:	Without Wind	With Wind	Net
Purchased Energy Savings	—		
Purchased Demand Savings	—		
		$149,400	$149,400

Annual Costs:	Hydro	Wind	Trans.	With Wind	Net
Operation and Maintenance	—	$22,500	$1,100		
Property Tax	—	19,014	2,200		
Depreciation	—	31,944	3,696		
	—	$73,458	$6,996	80,454	80,454
				$ 69,946	$ 68,946

Net Income:	
Bond Interest at 3.25%	13,789
Taxable Income	$ 55,157
Federal Income Tax (25%)	13,789
Available for Dividends at 6.5%	$ 41,368

TABLE XX. COMPARATIVE FINANCIAL STATEMENTS FOR 9000 KILOWATT WIND-POWER DEVELOPMENT

(Assuming Purchased Energy at $0.0045/kwh.; No State Tax on Wind-Energy; No Associated Hydro Plant for Reducing Purchased Demand)

BALANCE SHEET

ASSETS	Without Wind	With Wind	Net
Hydro Plant Additions	—	$824,700	$373,900
Wind-Power Development	—	110,000	560,800
Transmission Line	—		
		$934,700	$934,700

LIABILITIES			Net
Bonds (40%)			$373,900
Stocks (60%)			560,800
			$934,700

INCOME STATEMENT

	Without Wind	With Wind			Net
	Hydro	Hydro	Wind	Trans.	
Revenue:					
Purchased Energy Savings	—				
Purchased Demand Savings	—				
		$134,460			$134,460
Annual Costs:					
Operation and Maintenance	—		$22,500	$1,100	
Property Tax	—		16,494	2,200	
Depreciation	—		27,710	3,696	
			$66,704	$6,996	73,700
Net Income:					
			73,700		
		$60,760			$60,760
Bond Interest at 3.25%					12,151
Taxable Income					$48,609
Federal Income Tax (25%)					12,152
Available for Dividends at 6.5%					$36,457

TABLE XXI. COMPARATIVE FINANCIAL STATEMENTS FOR 9000 KILOWATT WIND-POWER DEVELOPMENT
(Assuming Purchased Energy at $0.0050/kwh.; Demand at $27/kw.; No State Tax on Wind-Energy)

BALANCE SHEET

ASSETS	Without Wind	With Wind	Net	LIABILITIES	Net
Hydro Plant Additions	$358,800	$ 468,000	$ 109,200	Bonds (40%)	$ 560,700
Wind-Power Development	—	1,182,400	1,122,400	Stock (60%)	840,900
Transmission Line	—	110,000	110,000		
	$358,800	$1,760,400	$1,401,600		$1,401,600

INCOME STATEMENT

Revenue:	Without Wind	With Wind	Net
Purchased Energy Savings		$149,400	
Purchased Demand Savings	$57,400	97,200	
	$57,400	$246,600	$189,200

Annual Costs:	Hydro	Hydro	Wind	Trans.	Net
	Without Wind	*With Wind*			
Operation and Maintenance	$3,588	$4,680	$22,500	$1,100	
Property Tax	7,176	9,360	23,648	2,200	
Depreciation	6,389	8,330	39,729	3,696	
	17,151		115,243		98,092
	$40,249		$131,357		$ 91,108

Net Income:	Net
Bond Interest at 3.25%	18,225
Taxable Income	$72,885
Federal Income Tax (25%)	18,221
Available for Dividends at 6.5%	$54,664

TABLE XXII. COMPARATIVE FINANCIAL STATEMENTS FOR 9000 KILOWATT WIND-POWER DEVELOPMENT
(Assuming Purchased Energy at $0.0045/kwh.; Demand at $15/yr./kw.; No State Tax on Wind-Energy)

BALANCE SHEET

Assets	Without Wind	With Wind	Net	Liabilities	Net
Hydro Plant Additions	—	$ 468,000	$468,000	Bonds (40%)	$ 565,200
Wind-Power Development	—	834,900	834,900	Stock (60%)	847,700
Transmission Line	—	110,000	110,000		
		$1,412,900	$1,412,900		$1,412,900

INCOME STATEMENT

Revenue:	Without Wind	With Wind	Net
Purchased Energy Savings		$134,460	
Purchased Demand Savings		54,000	
	—	$188,460	$188,460

Annual Costs:	Hydro	Wind	Trans.	Net
Operation and Maintenance	$ 4,680	$22,500	$1,100	
Property Tax	9,360	16,698	2,200	
Depreciation	8,330	28,053	3,696	
	$22,370	$67,251	$6,996	$ 96,617	$96,617

Net Income:		
	$ 91,843	$ 91,843
Bond Interest at 3.25%		18,369
Taxable Income		$ 73,474
Federal Income Tax (25%)		18,368
Available for Dividends at 6.5%		$ 55,106

TABLE XXIII. SUMMARY COMPARISON OF WIND-POWER VALUES ON CENTRAL VERMONT PUBLIC SERVICE CORPORATION SYSTEM

COMPETITIVE CONDITIONS								
Purchased Power Contract:								
Net energy charge (per kwh.)		$.0050				$.0045		
Cost of "demand" (per yr. per kw.)		$27				$15		

INVESTMENTS WARRANTED BY VALUE OF WIND	Total Value $	$/kw. of wind	Energy Value Only $	$/kw. of wind	Total Value $	$/kw. of wind	Energy Value Only $	$/kw. of wind
Total investment warranted	1,401,600		1,060,700		1,412,900		934,700	
Less associated extra hydro plant	109,200				468,000			
Warranted for wind power including transmission line	1,292,400	143.60	1,060,700	117.85	944,900	104.99	934,700	103.86
Add $4/kw. for reactive kva. value		4.00		4.00		4.00		4.00
Total warranted for wind-power and transmission line including reactive kva. value		147.60		121.85		108.99		107.86

197

Worth Versus Cost

Based on fixed and operating costs, as tabulated in the previous section, the total warranted investment by the Central Vermont Public Service Corporation for a 9000-kilowatt wind-power development, including 22 miles of transmission line, apparently lies between a maximum of $148 and a minimum of $108 per kilowatt of rated wind-generated capacity, with $125 as a likely value.

The higher figure also presupposes that Central Vermont Public Service Corporation, after further study, would find it practicable to embark upon 8600 kilowatts of additions to their hydro plants for the purpose of backing up the wind.

However, the cost of this 9000-kilowatt wind-power development, as determined by S. Morgan Smith Company in October, 1945, was about $205 a kilowatt. The only valid conclusion was that such a wind-power plant was not economically justified. Means of reducing the installed cost are explored in the next chapter. As fuel prices continue to rise, it seems likely that the gap could be bridged and that wind-turbines could be developed into additional sources of revenue for this and other utility systems.

SUMMARY

The costs of large-scale wind-power installations were determined, in 1945, based on a single hypothetical production run of 20 units rather similar in design to the test unit, to be about $190 a kilowatt installed at the switchboard, exclusive of transformers and connecting transmission line, and about $205 a kilowatt at the point of connection with the existing high-line.

The worth of wind-power was evaluated by Jackson and Moreland, who concluded that Central Vermont Public Service Corporation, for example, could afford to pay about $125 a kilowatt for a block of 9000 kilowatts.

Means of bridging this $80 gap are explored in the next chapter.

Chapter XII

WAYS TO REDUCE THE COST OF WIND–POWER

The cost, in October, 1945, of installing a block of 9000 kilowatts of wind capacity on Lincoln Ridge in Vermont, is reported at $191.11 a kilowatt exclusive of transformers and connecting high-line. This block of capacity would consist of six 1500-kilowatt units, weighing 497 pounds per kilowatt including the tower and foundation steel.

The possibilities for reducing this cost fall into six categories:

1. Cost reductions in the 1945 design by means of competitive bidding.
2. Cost reductions by means of refinements in the 1945 design.
3. Cost reductions by means of major modifications of the 1945 design.
4. Cost reductions by means of radical departures from the 1945 design.
5. Cost reductions resulting from the quantity production necessary to support a national wind-power program.
6. Cost reductions inherent in technological development.

The validity of estimating cost reductions under these six headings, item by item, and then adding them up to arrive at a possible total cost reduction, is open to a good deal of question. Some of my associates think it inevitable that the sum total of the savings described in this chapter will be realized and exceeded. Other associates doubt this. I shall itemize the possibilities, and leave it to the reader to make his own evaluation.

1. Cost Reductions in the 1945 Design by Means of Competitive Bidding

As explained in Chapter X, a large wind-turbine is made up partly of items which are in standard production and partly of items peculiar to the wind-turbine.

Reference is made in Chapter XI to quotations received from various suppliers from items in standard production. In no way should these figures be considered as competitive bids. In the first place, our specifications were not, in general, rigid enough to permit rigorous competitive bidding, with the result that each manufacturer modified the specifications somewhat according to his best judgment of our requirements. In the second place, in any such preliminary proposal engineering, there is inevitably present more or less of a cushion to provide a factor of safety against the uncertainties in the specifications and in general to provide for contingency. Thirdly, the estimates were requested in lots of 1, 5, 10 and 20 units, but it was not possible to specify the production schedule. Lastly, it was understood by the vendors that the quotations were for information only and were not final competitive bids.

In our judgment, true competitive bidding based on shop drawings detailed for a definite production schedule would have shown a cost reduction of not less than 5 per cent of the total cost of these standard production items.

As regards the second class of material entering into the Smith-Putnam Wind-Turbine, that is, items peculiar to the turbine, we lack even competitive estimates. The costs are based only on estimates prepared by the S. Morgan Smith Company. There is no reason to question the efficiency of the production engineering or the accuracy of the cost accounting of the S. Morgan Smith Company, both of which are thought to compare favorably with the practices of other manufacturers of comparable equipment. However, this is not to say that the items peculiar to the wind-turbine cannot be manufactured more cheaply than estimated by the S. Morgan Smith Company. The farming out of sub-assemblies to small shops frequently results in lower pound prices, particularly if a shop is found where the addition of a little load is welcomed as a means of justifying a second or a third shift; or of stabilizing employment; or of distributing overhead.

Our consensus is that the reduction in cost of the parts peculiar to the turbine by such manufacturing methods would be not less than 5 per cent.

2. Cost Reduction by Means of Refinements in the 1945 Design

In Chapter IX the recomputation of the loadings was discussed. These are the loadings used as the basis for the weight estimates of the preproduction design, the cost of which is analyzed in Chapter X. It will be recalled that the 1940 test unit weighed 500 pounds per kilowatt, and the 1945 preproduction design weighed 497 pounds per kilowatt, despite much simplification and cleaning up.

There are two ways in which the weight of the unit, and also its cost, may be reduced through a modification in the loadings. The first may be called "Refinement by Successive Approximations" and the second "Refinement by Modifying the Functional Requirements."

Refinement by Successive Approximations.

It has been pointed out in Chapter X that the determination of the weight of a rotating structure is made by a process of successive approximations. Since gravity or other acceleration forces are at work, the weight of the rotating members contributes to the total stresses. One assumes a distribution of weight, and runs through the calculations. If there is found to be an excess of weight, some may be removed in the 2nd approximation, bearing in mind that its removal will reduce some of the forces, and thus justify the removal of a little more.

It had been intended to carry through enough approximations of the preproduction unit to assure a cleaned-up model, designed to consistent strength. This has not been done. If done, it seems likely that there would be a weight reduction of not less than 7.5 per cent.

Refinements in the Loading Assumptions, by Modifying the Functional Requirements.

A review of the assumptions underlying certain of the loading conditions has revealed that acceptable alterations in operating and maintenance procedures would result in a reduction in the assumed maximum loading.

Maximum aerodynamic loading. For example, the most severe case of aero-dynamic loading discussed in Chapter X occurs when the blades are positioned vertically, and locked in both rotation and pitch, in the maximum wind. If this condition could be eliminated by a change in operating practice, which would leave the blades free to move in rotation and in pitch, then the maximum blade loading would be reduced by about 10 per cent. If a major repair does necessitate locking the blades vertically in periods of high wind, special supports and bracing can be rigged into the structure. There are other similar examples.

It is estimated that by these means the design stresses used in Chapter X could be reduced by 10 per cent, with some resultant reduction in weight.

Reduction in maximum torque by means of the electric coupling. A further small but real reduction in loading comes from a different source. When a synchronous generator is driven through an hydraulic coupling, the maximum torque that can be delivered by the turbine is limited only by the pull-out torque of the generator, that is, by the torque necessary to jump the generator out of phase with the transmission system with which it is interconnected. The pull-out torque of a generator is within the control of the designer. In practice on the Smith-Putnam Wind-Turbine, whenever the input torque reached 2.5 times the rated capacity of the generator, the generator was tripped off the line by a safety relay.

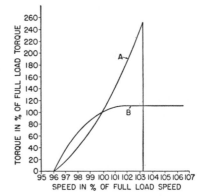

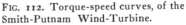

FIG. 112. Torque-speed curves, of the Smith-Putnam Wind-Turbine.

Curve A Using an hydraulic coupling

Curve B Using an electric coupling

However, for reasons explained in Chapter X, it is felt that any future large wind-turbine should not use an hydraulic coupling, but rather an electric coupling. The maximum torque transmitted by the electric coupling is in the control of the designer. If the specified maximum torque is exceeded, the generator is not affected, the excess torque merely causing an increased slip in the coupling. Thus, it is possible with an electric coupling to specify that the maximum input torque will not exceed 1.10 times the full-load torque. Fig. 112

shows the torque-speed curves for a Smith-Putnam Wind-Turbine with an hydraulic and an electric coupling, respectively. The reduction in maximum input torque from 2.5 to 1.10 times the full-load torque reduces the maximum stresses in the structure by an amount which it is not easy to evaluate. Our best estimate is that this stress reduction would be reflected in a reduction in the total weight of the unit of about 2.5 per cent.

Refinements in Loading Assumptions on Minor Members.

Finally, it seems likely that a complete redesign campaign would secure as one of its dividends similar refinements of the loadings on various of the minor parts of the structure. It is too much to expect that a complete illumination of this problem would decrease *all* the loadings, but it seems equally improbable that the net result would not be an over-all reduction in weight.

Our consensus is that the total cost of the unit might be reduced 15 per cent by refinements in the loading assumptions.

3. Cost Reductions by Means of Major Modifications of the 1945 Design

Flaps.

One of the most attractive modifications would seem to be the use of controllable flaps, instead of controllable pitch, for the regulation of power input.

The rotational speed and the output of the test unit were controlled by changing the pitch angle of the blades. Fig. 66 shows schematic layouts of several types of flap to accomplish the same end. If flaps proved satisfactory aerodynamically, their use would result in very substantial structural advantages.

For example, the connection of the blade to the A-frame becomes not only a simple problem, since there is no need for relative motion between the two parts, but, also, a region of high stress concentration in moving parts is replaced by a simple structure with much lower stresses.

Among the expensive items eliminated by this arrangement are the shank forging and its costly supporting bearings. It is true that the operating mechanism becomes more complex, but the individual parts become much smaller so that the mechanism to operate the flaps should cost less and require less servo power than the mechanism to operate pitch control.

Von Kármán raised the question of flaps in 1939, but no definitive study has been made to my knowledge to determine whether control of the turbine under all conditions would be satisfactory with flaps. Today, we are in about the same position as regards flap control as we were in regarding pitch control in 1939.

We think the cost of the unit would be reduced about $25 per kilowatt by the use of flaps.

Elimination of Coning Damping.

Another modification which should simplify the unit somewhat is the elimination of the coning-damping mechanism. The blades would then cone freely. Provision for coning-damping was built into the test unit and the initial runs were made with the coning heavily dampened. Operation was rough. The coning damping was relaxed and the operation became progressively smoother; when the coning damping was virtually eliminated, the operation was at its smoothest. We lack test data adequate to settle the point whether there should be a small amount of coning damping or whether the blades could be designed to cone freely without paying any important penalty in the operation of the unit. If the latter should prove to be the case, there would be a saving of about $0.50 per kilowatt.

Elimination of Yaw-Damping.

Another modification in the direction of simplification would be the elimination of the yaw-damping mechanism. This device had been introduced into the redesigned test unit as a modification calculated to smooth out the operation of the unit in yaw. However, the engineers felt that, although this device was included in the cost schedules of Chapter XI, it was perhaps not absolutely essential, even on the test unit as designed. And the reduction of e to zero (page 159), and the decrease in the overhang of the plane of rotation, both incorporated in Dornbirer's plans for the preproduction unit, would go so far toward eliminating the residual roughness in the operation of the test unit as to make it possible to eliminate the yaw-damping mechanism and to substitute for it a direct connected yaw-mechanism similar to that installed on the test unit. The substitution of this simpler mechanism should reduce the cost of the unit about $1 per kilowatt.

Skin-Stressed House.

There is another modification which is probably in the direction of cleaner structural design and which might result in a small reduction in cost. In the design of Chapter X, a stiff, box-type pintle girder supported all the machinery aloft, which was protected from the weather by wrapping a light sheet steel house around the whole thing. In a production design one might make the house and pintle girder into one unit. The house would be at least partially skin-stressed and, therefore, something more than self-supporting; such a structure would probably show an appreciable saving in weight but might cost more per pound, especially in small production runs. Arbitrarily, therefore, we estimate that by combining these two structures we would save $0.50 per kilowatt.

Conical Sheet Tower.

A further modification was partially explored in 1945. I refer to a conical tower constructed of sheet steel and looking like an ice-cream cone upside down.

The greatest advantage of this type of tower compared with the four-legged tower was found to occur where the foundation rock was sound and the tower could be secured by anchor bolts directly imbedded in the rock, eliminating the necessity for an expensive foundation excavation, of the sort we had to make at Grandpa's Knob. Costs of this type of tower were obtained from the Chicago Bridge and Iron Company, based on a somewhat incomplete stress analysis, whose uncertainties were guarded against by using plate thicknesses which were probably a little heavier than would be selected after a rigorous design. Ignoring this factor, however, it was found that in lots of 20 and where the foundation conditions were good, the use of the conical tower design would result in a saving of only about $0.15 per kilowatt; where the foundation conditions required excavation there would be no saving.

Conforming to the Most Economical Dimensions.

We have now made four determinations of the most economical dimensions of a large wind-turbine, with increasing confidence in the results. The dimensions used in the cost analysis of Chapter X were substantially those of the test unit which, it will be remembered, had been deliberately selected as being within range of the most economical dimensions, but on the small side.

We think that by conforming to the most economical dimensions, as determined by a fifth approximation, there would be realized a reduction in unit cost of about $1 per kilowatt.

Recapitulation of Possible Cost Reductions by Means of Major Modifications in the 1945 Design:

Modification	Cost Reduction
1. Flap control	$25.00/kw.
2. Elimination of coning-damping	0.50/kw.
3. Elimination of yaw-damping	1.00/kw.
4. Combined house and pintle girder	0.50/kw.
5. Shell-type tower	—
6. Optimum dimensions	1.00/kw.
Total saving	$28.00/kw.

Recapitulation of the Total Cost Reductions Which Might Be Realized by Competitive Engineering and Bidding, Refinements in the Loading Assumptions of the 1945 Design and Major Modifications of the 1945 Design:

Estimated installed cost at the switchboard on Lincoln Ridge, Vermont, of six 1500-kilowatt units, diameter 175 feet, turbine speed 31.5 revolutions per minute, generator speed 900 revolutions per minute, tower height 150 feet—August, 1945 .... $191.11/kw.

WAYS TO REDUCE THE COST OF WIND-POWER

Cost reductions by means of:

a. Competitive bidding in the 1945 design, 5% ...... $ 9.55
b. Refinements in the 1945 design, 15% 27.22
c. Major modification in the 1945 design 28.00

Total cost reduction $64.77

Net cost .... $126.34/kw.

4. Cost Reduction by Means of Radical Departures from the 1945 Design

Elimination of Gear and Coupling.

The gear is the bottleneck in the design of a large wind-turbine. Many schemes have been proposed looking to its elimination. One of these, made by Honnef of Berlin, is described in Chapter VI, together with reasons why that particular suggestion was considered impractical.

It might be possible to connect a 2000-kilowatt generator directly to a large wind-turbine with a rotational speed of about 30 revolutions per minute. The construction of such a generator is practical if the members are sufficiently rigid to retain the air-gap within the proper limits. The elimination of the gear and other equipment would reduce the cost by about $21 per kilowatt. Development studies would be required to determine if the increase in cost of the special low-speed generator would be less than this.

Combination of Generator and Electric Coupling in One Unit.

A less drastic suggestion is the combination of a conventional generator and an electric coupling into one unit. Such a composite design has been developed on an experimental basis. Whether it could be made satisfactory in wind-turbine service and what the cost per kilowatt would be in limited mass production can only be determined by further study.

Elimination of Coning.

A structural suggestion frequently repeated is that the blades should be fixed at the hub with no freedom to cone. It is possible that definitive cost studies of designs with and without coning would show that a design with three fixed blades would prove economically superior to one with three blades free to cone. Our dynamic studies have indicated that for the two-bladed design coning is better, and it is questionable whether there would be any economic gain in going to three fixed blades.

There are still other possibilities, too speculative to be mentioned here, and it is futile even to guess at the economic gains, if any, to be achieved by this sort of thinking.

5. Cost Reductions Resulting from Quantity Production Necessary to Support a National Wind-Power Program

The costs of Chapter XI were based on a total production of twenty 1500-kilowatt units, and the pound price quoted by S. Morgan Smith Company for the 30,000-pound mild-steel blades, for example, was $0.44 a pound. The over-all manufacturing pound price for the unit was $0.39 a pound.

Some comments are necessary to complete the assessment of such cost estimates.

The design had not been analyzed for production. The costs had been estimated in lots of 1 and 20. The percentage reduction in the unit cost of a lot of twenty blades, hub assemblies, and pintle assemblies was 9 per cent, 10 per cent, and 7 per cent, respectively, as compared with the estimated costs of lots of 1. What this reduction amounts to when elaborate jigging and fixtures can be used to advantage is seen in Fig. 85, which shows that the unit cost of twenty pairs of stainless-steel blades is estimated by the Budd Company to be only about 50 per cent of the cost of one pair.

If a national program were to get under way, in Scotland, New Zealand, or the United States, the price of $0.44 per pound for the blades would tend to shift toward the price of $0.06 per pound at which 45,000-pound mild-steel box cars and gondolas were being sold in August, 1945. It is true that the car builders were able to offer such a price because their shops were tooled for a production of tens of thousands of freight cars a year, and it is equally true that even national wind-power programs would hardly consume more than a few thousand 2000-kilowatt or 3000-kilowatt wind-turbines over a period of years. Still, this would be something better than a total production run of 20 units, and the over-all competitive pound price of a design engineered for production in lots of 100 per year would be something less than $0.39, by perhaps 15 per cent, bringing the over-all pound price to $0.33 (1945 prices).

6. Cost Reductions Inherent in Technological Progress.

It is a truism that the history of many technical developments is one of relatively decreasing unit costs. Today we can buy for $1500 a better car than we could get for $5000 in 1915. So it is also with household refrigerators, radios, frozen foods, and many products where competition has resulted in aggressive engineering.

It is very unlikely that the speculations of this chapter have exhausted the possibilities which lie in the future of the large-scale wind-turbine. After all, our experience is limited to 1100 hours of operation of a single test unit, not designed with an eye to low production costs.

It seems inevitable that the hundred and first production unit, and even the eleventh, would contain many refinements contributing to lowered unit costs not foreseen in this chapter.

WAYS TO REDUCE THE COST OF WIND-POWER

SUMMARY

In Chapter XI it was found that a preproduction unit closely following the test unit in design would cost $191.11 a kilowatt installed on Lincoln Ridge in Vermont, exclusive of transformers and connecting high-line. Six means of reducing this cost are considered. Estimates of cost reduction by each of these means are offered. Some of these estimates rest on good evidence; some are highly speculative. If all of the suggestions could be realized and if no unfavorable factors remain to be discovered, it is possible that 2000-kilowatt or 3000-kilowatt windturbines, produced in quantities sufficient to support a national wind-power program, could be installed for about $100 a kilowatt, exclusive of transformers and connecting high-line (1945 prices).

Chapter XIII

THE FUTURE OF WIND-POWER

J. B. S. Haldane, in *The Last Judgement,* records verbatim a Children's Hour broadcast from Venus in the year 40 million. In describing the end of human life on Earth, the commentator remarks: "It was characteristic of the dwellers on Earth that they never looked ahead more than a million years, and the amount of energy available was ridiculously squandered."

Oil. It is true that our reserves of oil contain no more heat than we receive from the Sun in the course of a day or so, and that we are burning them up with the abandon of spendthrifts. The United States, once the greatest producer, now imports oil.

Coal. It is also true that one of the world's great industrial countries, Great Britain, has run short of coal. The mines are old, the working faces are far from the pit-heads, and the miners are telling their sons to quit mining. In the winter of 1946–1947 Britain, who exported as much as 40,000,000 tons of coal annually before the war, exported virtually none. Indeed, she did not have enough for herself.

And in this country it seems likely that coal prices will go higher.

Nuclear energy has been brilliantly harnessed. Whether the Atomic Energy Commission will feel free in the years just ahead to release uranium for general use in power piles remains to be seen. Without much to go on, it has been estimated that a 100,000-kilowatt nuclear power plant would cost about $270 per kilowatt and would break even with $9 coal. And it has been pointed out that the general application of nuclear fuels would probably be confined to the very large central stations. Oppenheimer, testifying at Lake Success as a guest witness before the Atomic Energy Commission of the United Nations in June, 1947, estimated that the application of atomic fuels to large central stations lay twenty to thirty years in the future.

On July 24, 1947, the *New York Times* quoted the U. S. Atomic Energy Commission as saying, in its first semiannual report: "But a number of basic advances in physics, chemistry and metallurgy will be required before power is produced at satisfactory efficiency and cost. The technical problems to be overcome are many, but we confidently expect them to be solved."

In any event, such substitution of nuclear energy for coal as the fuel in central stations would not appear to have more than incidental bearing on the future of wind-power.

The sun. The power available in the direct radiation from the sun is colossal, but only in deserts in low latitudes is it possible to extract some of this power economically.

The tides. Widespread extraction of power from the tides remains an economic fantasy.

Wind-power. It is hardly possible to make a tally of the world's available wind-power as one does of water-power. We know the world stream-flow in cubic feet per second fairly accurately and annual rainfall gives us a useful cross check. We know the average height through which this mass of water falls in getting to the sea or to the land-locked lakes. The product, in foot-pounds per second or horse-power, is a matter of arithmetic. Thus, the world's potential water-power amounts to about 5×10^8 kilowatts. Of this, about 5×10^7, or 10 per cent, has been developed.

The case of wind is different. True, we can weigh the atmospheric envelope and assume a mean speed for it. Brunt (Ref. 12-A), who has done this, estimates the total power in the atmosphere to be 3×10^{17} kilowatts. But not all of this power is available to man—only that portion in the lowest stratum of the atmosphere.

Willett,* after making certain assumptions, concludes that the wind-power available to wind-turbines amounts to about 2×10^{10} kilowatts. Only national wind-power surveys can determine what percentage of this 20 billion kilowatts of wind-power is to be found on the type of site described in Chapter IV, and near load centers.

What is the potential role for some of this 20 billions of kilowatts of wind-power, especially in conjunction with water and fuels, including nuclear fuels?

The Extent of the Market

In Vermont, we have found a good wind-power site—Lincoln Ridge—with a capacity for some 50,000 kilowatts. Like a good site for a large dam, a good site for a large block of wind-power is a topographic rarity. To provoke discussion, I will guess that somewhere on earth there are 49 more sites like Lincoln Ridge, close to heavy load centers. Most of these sites would be found in such windy, industrialized regions as Scotland, northern Ireland, Iceland, Newfoundland, the Maritime Provinces of Canada, New England, other parts of the United States, southern Chile, New Zealand, Tasmania, and possibly high in the Italian Apennines, and in Scandinavia and other regions itemized in Tables I and II in Chapter II. In the aggregate these 50 sites would amount to a potential market for large wind-turbines of 2,500,000 kilowatts.

* Private communication.

If governments should sponsor wind-power projects, using a standard design of about 2500 kilowatts in order to obtain the benefits of mass production and standard maintenance, then conceivably 100 such units, aggregating about 250,000 kilowatts, might be installed annually for a number of years.

At the other extreme lies the wind-power set, rated at 1 kilowatt or less, used for farm lighting and radio battery charging in districts without rural electrification. About 10,000 are being sold each year by United States manufacturers. If it is assumed that this market will not be saturated for twenty years, it will have amounted to 200,000 kilowatts by 1967.

Between these two extremes much interest has been shown in wind-power plants of a few hundred kilowatts or less.

In the period 1940–1946, the S. Morgan Smith Company, without having advertised the experiment on Grandpa's Knob, received hundreds of inquiries from all quarters of the globe for wind-power plants rated at about 100 kilowatts, for isolated use; or use in conjunction with hydro alone; or with fuel generation alone; or on small combined water and fuel systems. Although one or two experimental plants of about this size have been built, as described in Chapter VI, such a plant has never been designed for production and the market for it has never been estimated.

The Worth of Wind-Power Installations

Table XXII (Chapter XI) shows how the worth of wind-power varies with the size of the installation; with the windiness of the site; with the amount of the annual charges; and with the credit for capacity value. The 1-kilowatt set for farm lighting, when properly installed on a tower of suitable height, commands a market price of $300 to $600 a kilowatt. The 100-kilowatt set is worth from about $100 a kilowatt based on energy value alone to over $600 a kilowatt where a credit for capacity value can be earned. The 3000-kilowatt set similarly is worth from about $50 to over $300 a kilowatt.

The Installed Costs of Wind-Power Installations

The family of non-linear curves of Fig. 113 has been prepared to suggest how the 1945 installed cost per kilowatt would have varied with variation in the capacity in kilowatts, and with variation in the manufacturing rate. This family of curves has been developed around only two rather vague fixes—the known unit cost of the 1-kilowatt set in lots of 10,000 per year, and the estimated unit cost of the 1500-kilowatt set in lots of 6 per year—the whole embellished by the speculations of Chapter XII concerning the least cost of a 2000- or a 3000-kilowatt unit, in lots of 100 per year. It must be realized, therefore, that the curves have no serious quantitative meaning.

For comparison, the range of the worth of each of the three sizes, taken from

Table XVII, is also shown in Fig. 113 by means of three vertical arrows. The bull's-eyes indicate representative values.

Assuming that the 1-kilowatt set was being sold at a profit in 1945, then, if the speculations of Chapter XII have validity, it would have been possible in 1945 to manufacture and install the 2000- or 3000-kilowatt (and perhaps the 100-kilowatt) unit at a profit, provided that the market could have absorbed 100 or more units a year.

In the absence of both a national wind-power survey and cost studies based on production designs, it does not seem profitable to push these computations further. The crystal ball is not yet in focus. The conclusions to be drawn from the blurred image so dimly seen follow.

Conclusions Regarding the Future of Wind-Power

1. Nuclear energy is another source of heat, but its economics are not yet accurately known, and in the foreseeable future nuclear energy will have no effect on the market for large-scale wind-power.

2. In the foreseeable future, nuclear energy will not displace water-power.

3. Neither solar nor tidal-power will be harnessed on a large scale in the near future.

4. As long as water-power remains economically justified, special partnerships between wind and water will be justified.

5. Coal prices will continue to increase.

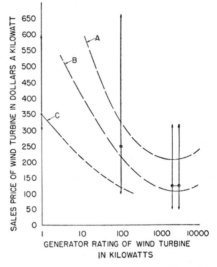

FIG. 113. First approximation of the variation in the 1945 worth of a wind-turbine with variation in the rating, for three assumed annual manufacturing rates.

Curve A   6 units per year
Curve B   100 pnits per year
Curve C   10,000 units per year

Vertical arrows show ranges of worth taken from Table XVII. Double circles indicate most probable values.

6. The market for wind-power plants may fall into four groups, characterized by the size of the unit:

A. The largest unit would be rated at 2000 or 3000 kilowatts, for addition to existing power systems, principally in conventional support of water and steam. This application is limited to those selected sites near heavy load centers which occur most frequently in windy regions between latitudes 30 and 60 degrees, North and South. The market may range in size from 1,000,000 kilowatts to 10,000,000 kilowatts.

B. The medium unit would be rated at 100 to 500 kilowatts, for use in conjunction with small hydroelectric installations, or Diesel sets, in windy, isolated communities, such as the Shetlands, the Orkneys and some of the islands in the trade winds. The market may range in size from 250,000 kilowatts to 2,500,000 kilowatts.

C. A small unit of about 10 kilowatts would have a special limited use in charging batteries for untended airway beacons, as in the Arctic, and perhaps some desert regions. The market may range from 1000 kilowatts to 10,000 kilowatts.

D. The smallest unit of 1 kilowatt or less, is for farm lighting. The market may range in size from 250,000 kilowatts to 2,500,000 kilowatts.

7. The range in the worth, measured in dollars per kilowatt, at the 1-, the 100-, and the 3000-kilowatt sizes, respectively, is discussed in Chapter XI, tabulated in Table XVI, and compared with very uncertain estimates of manufacturing costs in Fig. 113, which indicates that production runs of about 100 units a year would permit selling at a profit in the 100- and the 3000-kilowatt markets.

8. Possible applications for intermittent wind-energy, in isolated packages, are listed in Chapter XI.

9. The first step toward harnessing wind-power in blocks larger than a few kilowatts should be a national wind-power survey, the specifications for which are discussed in Chapter V.

10. Grandpa's Knob has demonstrated that the technical problems of the 1250-kilowatt wind-turbine are understood and have been solved. To solve the economic problems of putting this or a larger wind-turbine into low-cost production probably requires Government aid.

BIBLIOGRAPHY A—GENERAL REFERENCES*

1. Baker, O. E.
   Atlas of American Agriculture, Government Printing Office—1936
2. Sherlock, R. H. and Stout, M. B.
   Relation Between Wind Velocity and Height During a Winter Storm
3. Wells, B. W. and Shunk, I. V.
   Salt Spray—An Important Factor in Coastal Ecology, *Bulletin* Torrey Botanical Club 65, 485–492—October 1938
4. Willhofft, F. O.
   Industrial Applications of the Flettner Rotor, *Mech. Eng.*, Vol. 49—No. 3
5. *The Electrician*, Electricity from the Wind—November 24, 1933
   Electrical World, Wind Rotor Experiments "Decidedly Satisfactory"—October 29, 1933
6. Savonius, S. J.
   The S Rotor and Its Applications, *Mech. Eng.*—May 1931
   Klemin, Alexander
   The Savonius Wind Rotor, *Mech. Eng.*—November 1925
7. *Scientific American*, Aerodynamic Windmills—June 1929
8. Darrieus
   Les Moteurs a Vent. Les Colines Electriques, *La Nature*—December 15, 1929
9. Sectorov, W. R.
   Report on the Operating Characteristics of the Initial 100 KW Aero-electric Unit at Balaklava (Translated), *Elektrichestvo*—No. 2 of 1933
10. *The Electrician*, Electricity from Wind Power—December 8, 1933
11. Glauert, H.
    Anscrew Theory—Vol. IV, Section L of *Aerodynamic Theory*, edited by Durand
12. Brunt, D.
    Physical and Dynamical Meteorology, Cambridge University Press—1934

BIBLIOGRAPHY B

A partially annotated bibliography of upwards of a thousand references dealing with the prior art, in English, French, German, Italian, Spanish, and Russian, is available at the York, Pennsylvania, offices of the S. Morgan Smith Company.

The following references are to be found only in the S. Morgan Smith Company files:

1. Petterssen, Dr. Sverre
   Wind Regimes of the World, Preliminary Report—April 15, 1940
   Second Report—August 9, 1940
   Appendix I to Reports—December 10, 1940

*Reference numbers followed by an A in the text, such as 1-A, are general references, whereas those followed by a B are to be found in the S. Morgan Smith Company files only.

BIBLIOGRAPHY

2. Petterssen, Dr. Sverre
   Preliminary Report on the Energy of the Winds in the New England Area—
   April 1940
   Report on Additional Research in Connection with the Energy of the Winds
   in the New England Area—April 1940
   Reports on Wind Observations in Vermont—July 1–October 23, 1940
   Appendix I to Report on Wind Observations in Vermont—January 20, 1941
   Summary Report on Site Investigations—June 10, 1940
   Appendix II to Report on Wind Observations in Vermont—March 14, 1941
   Site Specifications for Test Site and Memoranda on Site Specifications for the
   Akron Symposium—July 29, 1940
3. von Kármán, Dr. Th.
   Preliminary Report on the Aerodynamic Characteristics of the Smith-Putnam
   Wind-Turbine—January 5, 1940
   Second Report on the Aerodynamics of the Smith-Putnam Wind-Turbine—1940
4. Wilcox, Carl J.
   Memorandum on the Computation of Mean Annual Weighted Density—
   April 1941
5. Wilcox, Carl J.
   Computation of Mean Annual Weighted Density at Sea Level, Blue Hill,
   Mt. Abraham, Lincoln Mt., Mt. Washington and 10,000 Ft.—September 1945
6. Brooks, Dr. Charles F.
   Construction of Maps of Maximum Icing and Maximum Wind for the U. S.,
   Southern Canada and the South Coast of Alaska—January 6, 1941 (Bound
   with Appendix II to Report III by S. Petterssen)
7. Griggs, Dr. Robert F.
   Reports on Field Trips to Central Vermont—May 25, 29, 1940, and June 30,
   1940
8. Troller, Dr. Theodore
   S. Morgan Smith Wind-Turbine Site Tests, Pond Mt.—April 1940
   Further Report on Pond Mt.—June 1940
   East and Glastenbury Mts.—April 1940
   Mt. Washington—April 1940
   Summary Report—June 1940
9. Wilcox, Carl J.
   Anomaly Study—April 1941
   Petterssen, Dr. Sverre
   Appendix I to Report on Wind Observations in Vermont—January 20, 1941
10. Putnam, P. C.
    Predictable Wind Power—June 14, 1939
11. Lange, Dr. Karl O.
    Anemometry Suggestions and Recommendation—March 1940
12. Harvard University Aerodynamics Laboratory
    Report No. 102—March 15, 1940—William Bollay
    Report No. 103—November 1940—A. E. Puckett
    Report No. 104—February 25, 1941—A. E. Puckett
    Report No. 105—no date—A. E. Puckett
    Report No. 106—April 28, 1941—A. E. Puckett
13. Wilcox, Carl J.
    Handbook of Aerology, Vol. III, Site Factor and Variation with Height Com-
    putations Test Site

Bibliography

    Memorandum on the Computation of Variation with Height—February 21, 1941

    Mt. Washington with Correction Factors for Anemometer Locations—September 1945

    Original Data—Comparison of Old and New Masts

    Miscellaneous Correspondence—Original Data for Report

    Original Data for Grandpa's Knob by Months, Southwest Winds Only

    Report on Vertical Velocity Gradients on Grandpa's Knob—December 1941

14. Griggs, Dr. Robert F.

    Tree Reactions to Wind—February 14, 1940

    Report on Field Trips to Central Vermont—May 1940

    Report on Examination of Mount Ellen Ridge—March 1, 1945

    Report on Field Trip—May 26 through June 4, 1945

15. Wilcox, Carl J.

    Report on Vertical Velocity Gradients on Grandpa's Knob—December 1941

16. Wilcox, Carl J.

    Computations on Variation of Velocity with Height, Vol. III, Handbook of Aerology—1941

17. Wilcox, Carl J.

    Mt. Washington Gradient with Correction Factors for Anemometer Locations—September 1945

18. Wilcox, Carl J. and Dornbirer, S. D.

    Large-Scale Wind Power Analysis—October 1945

19. Wilcox, Carl J.

    Static and Dynamic Characteristics of the Smith-Putnam Wind-Turbine—1940

    Memorandum on the Computation of Outputs—February 1941

20. Wilcox, Carl J.

    Output Summaries—1940–1945, Vol. IV, Handbook of Aerology

    Anomaly Study—April 1941–December 1946

21. Wilcox, Carl J.

    Memorandum on Diurnal Variation of Output—February 18, 1944

22. Voaden, Grant H.

    Field Test Report No. 1—Preliminary Report of Field Tests on Wind-Turbine Speed Regulation—November 14, 1941

    Smith-Putnam Wind-Turbine Field Test Report No. 2—December 10, 1941

    Smith-Putnam Wind-Turbine Field Test Report No. 3—February 2, 1942

    Smith-Putnam Wind-Turbine Field Test Report No. 4—March 18, 1942

    Smith-Putnam Wind-Turbine Field Test Report No. 5—March 26, 1942

    Smith-Putnam Wind-Turbine Field Test Report No. 6—April 4, 1942

    Smith-Putnam Wind-Turbine Field Test Report No. 7—May 20, 1942

    Smith-Putnam Wind-Turbine Field Test Report No. 8—September 21, 1942

    Smith-Putnam Wind-Turbine Field Test Report No. 9—January 14, 1943

    Smith-Putnam Wind-Turbine Field Test Report No. 10—April 1, 1943

23. Haynes, S. S.

    Comparison of Forecasts—April 29, 1942

    Reviewed and brought up to date by G. H. Cheney—February 1944

24. Putnam, P. C.

    Report to Board of Directors, S. Morgan Smith Company—March 1940

25. Wilbur, Dr. J. B.

    General Specifications for Smith-Putnam Wind-Turbine—April 8, 1941

26. Voaden, G. H.

    Book of Instructions for Smith-Putnam Wind-Turbine—2 Vols.

27. Wilbur, Dr. J. B.
    Test Program Smith-Putnam Wind-Turbine Test Unit—May 10, 1941
28. Holley, M. J.
    Operating Instructions for Foxboro Recorder—June 11, 1941
29. Wilcox, Carl J.
    Evaluation of the First Run of the Turbine—October 19, 1941
30. Wilcox, Carl J. and Holley, M. J.
    Memo on Coning Prepared for von Kármán—July 1942
31. Test Engineers and Operators
    Log Sheets Daily—September 4, 1941–March 26, 1945
32. von Kármán, Dr. Th.
    Report on Aerodynamic Data for Choice of the Optimum Design—February 5,
    1940
33. Reid, Elliot G.
    Model Tests Smith-Putnam Wind-Turbine—February 3, 1941
34. Wilcox, Carl J.
    a. Outputs and Frequency Distribution Curves, New England Sites, Vol. VII,
       Handbook of Aerology—Eastern U. S.
    b. Outputs and Frequency Distribution Curves, New England Sites, Vol. XII,
       Handbook of Aerology
    c. Outputs and Frequency Distribution Curves for Oceanic Islands and the
       Maritime Littorals, Vol. XIII, Handbook of Aerology
35. Wilcox, Carl J.
    Cost Study of Induction Generation—1943
    Synchronous Generation—Optimum Study—1943
    Summary of Energy Costs Induction and Synchronous Generation—1943
36. Wilcox, Carl J.
    Report on Energy Costs—May 9, 1945
37. Wilcox, Carl J.
    Report on Choice of Blade Shape—1943
38. Rutland Office
    Basic Aerodynamic Calculations
39. Wilcox, Carl J. and Holley, M. J.
    Report on Yawing and Pitching Moments Transmitted to the Pintle Axis of
    the Smith-Putnam Wind-Turbine—April 1943
40. Wilcox, Carl J. and Holley, M. J.
    Determination of Design Loadings for Blades of Smith-Putnam Wind-Turbine—
    May 14, 1945
    Summary
    Reports on each of 10 Loading Conditions
41. Wilbur, Dr. J. B.; Wilcox, Carl J.; and Holley, M. J.
    Tentative Loading Specifications for the Blades of the Preproduction Model—
    December 22, 1943
42. U. S. Plywood Corporation
    Report on Preliminary Design of Plywood Blades for the Smith-Putnam Wind-
    Turbine—June 1, 1944
43. Wilcox, Carl J.; Holley, M. J.; and Dornbirer, S. D.
    Report on Cost Study of October 1945
44. Atkinson, K. (Jackson & Moreland)
    Preliminary Memorandum on Value of Wind Power—April 27, 1945

BIBLIOGRAPHY

45. Atkinson, K. (Jackson & Moreland)
    Recommended Draft for "The Worth of Wind Power to Central Vermont
    Public Service Corporation" (included in "Large-Scale Wind Power"—an
    analysis by C. J. Wilcox and S. D. Dornbirer, October 26, 1943)

INDEX